W0254338

Materials and Development of Plastics Packaging for the Consumer Market

Sheffield Packaging Technology

Series Editor: Geoff A. Giles, Worldwide Supply Operations, SmithKline Beecham, London.

A series which presents the current state of the art in chosen sectors of the packaging industry. Written at professional and reference level, it is directed at packaging technologists, those involved in the design and development of packaging, users of packaging and those who purchase packaging. The series will also be of interest to manufacturers of packaging machinery.

Titles in the series:

Design and Technology of Packaging Decoration for the Consumer Market
Edited by G.A. Giles

Materials and Development of Plastics Packaging for the Consumer Market
Edited by G.A. Giles and D.R. Bain

Technology of Plastics Packaging for the Consumer Market
Edited by G.A. Giles and D.R. Bain

Materials and Development of Plastics Packaging for the Consumer Market

Edited by

GEOFF A. GILES
Worldwide Supply Operations
SmithKline Beecham
London, UK

and

DAVID R. BAIN
Managing Director
Techplast Consulting Ltd
Wirral, UK

CRC Press

First published 2000

Published by
Sheffield Academic Press Ltd
Mansion House, 19 Kingfield Road
Sheffield S11 9AS, England

ISBN 1-84127-116-0

Published in the U.S.A. and Canada (only) by
CRC Press LLC
2000 Corporate Blvd., N.W.
Boca Raton, FL 33431, U.S.A.
Orders from the U.S.A. and Canada (only) to CRC Press LLC

U.S.A. and Canada only:
ISBN 0-8493-0507-1

Printed on acid-free paper in Great Britain by
Bookcraft Ltd, Midsomer Norton, Bath

British Library Cataloguing-in-Publication Data:
A catalogue record for this book is available from the British Library

Library of Congress Cataloging-in-Publication Data:
Materials and development of plastics packaging for the consumer market / edited by Geoff A. Giles, David R. Bain.
p. cm.--(Sheffield packaging technology; v. 2)
Includes bibliographical references and index.
ISBN 0-8493-0507-1 (alk. paper)
1. Plastics in packaging. I. Giles, Geoff A. II. Bain, David R.

TS198.3.P5 M38 2000
668.4'97--dc21

00-021385

Preface

Packaging makes an important contribution to all stages in the supply chain of delivering consumer products to the end user, and no material has made a greater contribution than plastics. From a limited number of high volume polymers, allied to a few variants with special properties such as barrier to oxygen or barrier to flavour loss, a wide range of packages can be made, ranging from flexible to rigid, to meet the diverse requirements of the marketplace. There are options to vary shape and colour, and a range of decorative techniques to increase further the marketing benefits.

This volume is designed to guide the reader through the technical and commercial issues arising from the use of plastics materials and the basic package types.

Plastics are lightweight and tough, and can be manufactured, filled and packed at high speed – contributing to reduced costs and the minimisation of packaging in the supply chain. In the retail outlets, plastics packaging is an essential element of the brand image, advertising the brand by attractive presentation. The product is protected and the quality is maintained until the pack is empty. Dispensing and reclosing systems provide consumer convenience. Tamper-evident and child-resistant closures contribute to safety. In many cases, plastics packaging has contributed greatly to the development and the brand image of familiar household products.

At the end of their lifecycle, plastics packs can be disposed of safely by a number of environmentally acceptable options—mechanical recycling, chemical recycling or incineration with energy recovery—the choice depending on the collection infrastructure and the facilities available. The objective is to use the most environmentally acceptable route, preferably supported by Life Cycle Analysis (LCA).

With the range of benefits available, it is not surprising that plastics are the materials of choice for many packaging applications. Consequently, despite lightweighting and other changes to achieve packaging minimisation, the total amount of plastics in the packaging market continues to grow.

This volume takes the reader through the opportunities and performance elements related to the use of plastics materials in packaging. It enables the reader, through better understanding, to make a more informed choice in the selection and use of the available materials.

As trends in developed markets could also be the direction for developing markets, an appendix on the trends in Western Europe has been included to provide an additional perspective.

In compiling this book, we have sought the support of chapter authors who are well experienced in their specialised areas. They are all busy people who have devoted valuable time to writing their chapters, and our sincere thanks go to them.

This volume is designed to provide a comprehensive source of reference to the range of materials available for plastics packaging. As such, it complements publications focussing on individual plastics materials.

Geoff Giles
David Bain

Contributors

Dr David Bain	Techplast Consulting Limited, 32 Long Meadow, Gayton, Wirral CH60 8QQ, UK
Mr David Brooks	Crown Cork & Seal Company, Inc., Downsview Road, Wantage, Oxon OX12 9BP, UK
Mr Christopher Cross	Marketpower, 84 Uxbridge Road, London W13 8RA, UK
Mr Ian Dent	BP Chemicals, Chertsey Road, Sudbury-on-Thames, Middlesex TW16 7LL, UK
Mr John Dixon	Lawson Mardon Packaging – Flexibles, PO Box 3, Midsomer Norton, Bath BA3 4AA, UK
Mr Geoff Giles	SmithKline Beecham, Three New Horizons Court, Brentford, Middlesex TW8 9EP, UK
Dr Harold Hughes	157 Packaging Building, Michigan State University, East Lansing, MI 48824-1223, USA
Mr Terry McCormack	Basell UK, Kingsfield Court, Chester Business Park, Chester CH4 9RE, UK
Mr Mike Naylon	Queensland Manufacturing Institute Limited, PO Box 4012, Eight Mile Plains, Queensland 4113, Australia
Mr Michael Rooney	Food Science Australia, 16 Julius Avenue, Riverside Corporate Park, Delhi Road, (PO Box 52) North Ryde NSW 2113, Australia
Mr Mark Shickle	PSD Associates Ltd, 16-18 Petersham Road, Richmond, Surrey TW10 6UW, UK
Mr Andrew Streeter	CPS International, 7 Market Place, Saffron Walden, Essex CB10 1HJ, UK

Contents

1 Technical and commercial considerations

D.R. Bain and G.A. Giles

1.1 Introduction

Plastics are essential and enduring materials for packaging. Despite the concerns of some environmental groups, their uses continue to grow, as their manifest benefits far outweigh any negative aspects—perceived or real. The number of polymers used in packaging is relatively small, consisting of high volume materials such as polyethylene (PE) in all its forms, polypropylene (PP), polystyrene (PS) and polyethylene terephthalate (PET), plus a few more specialised materials like ethylene-vinyl alcohol (EVOH), polyethylene naphthalate (PEN), nylon (PA) and polycarbonate (PC). Emerging materials, still to make an impact, include cyclic olefin copolymers (COC), styrene interpolymers and polylactic acid (PLA). Tonnages of plastics used in packaging in Western Europe are shown in Figures 1.1–1.6.

Most plastics materials are available globally from the major companies indicated in Tables 1.1 and 1.2, showing predicted global and European polyethylene capacities for 2000. These figures refer to total production, not just packaging. This ready availability of plastics materials around the world, as well as the

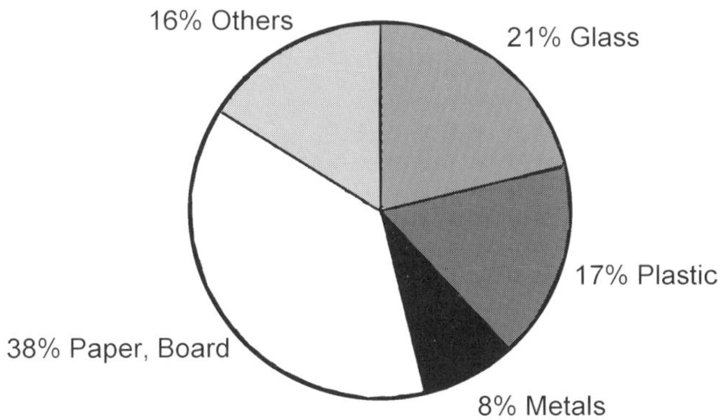

Figure 1.1 Average percentage of plastics packaging in total packaging (Western Europe, 1997). Percentages are expressed in weight of products. Data are the maximum and minimum percentage observed in Western Europe. Total: 66 400 706 tonnes, an increase of 1.4% in weight compared to 1996. Source: APME.

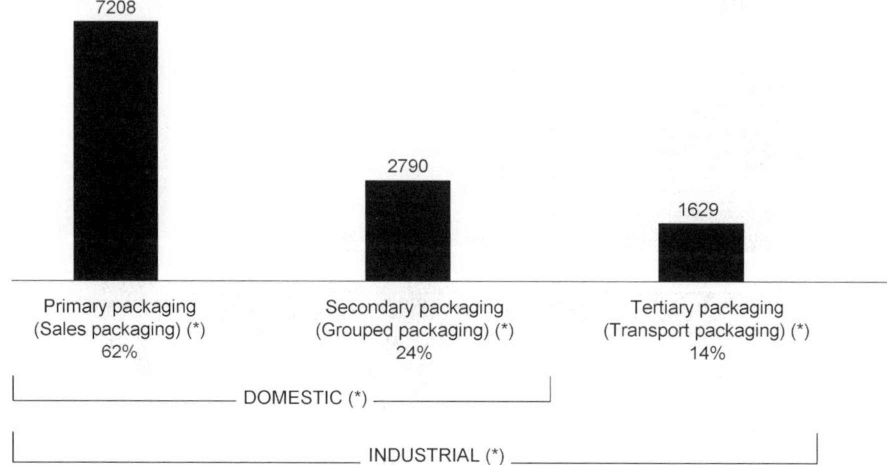

Figure 1.2 Plastics packaging by category (Western Europe, 1997), according to the European Directive on packaging(*). Total: 11 627 000 tonnes. Units: × 1000 tonnes/year. Source: APME.

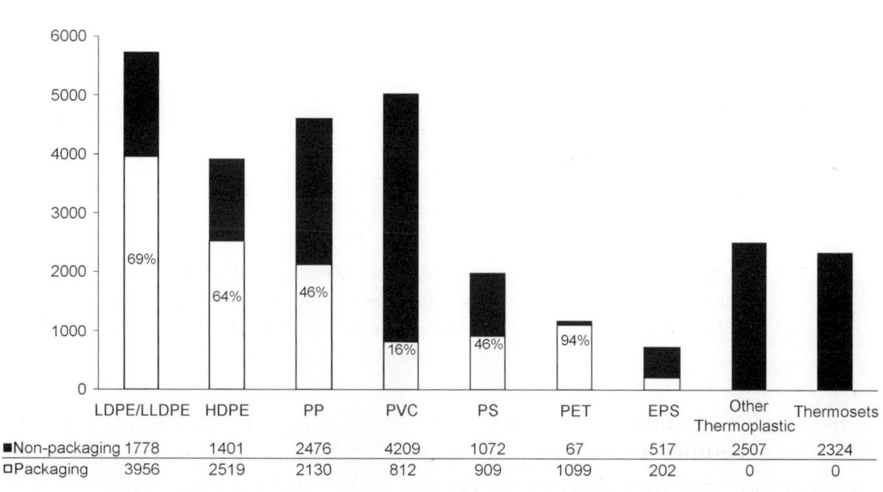

	LDPE/LLDPE	HDPE	PP	PVC	PS	PET	EPS	Other Thermoplastic	Thermosets
■Non-packaging	1778	1401	2476	4209	1072	67	517	2507	2324
□Packaging	3956	2519	2130	812	909	1099	202	0	0

Figure 1.3 Percentage of packaging by resin (Western Europe, 1997). Key: □ packaging; ■ non-packaging. Source: APME.

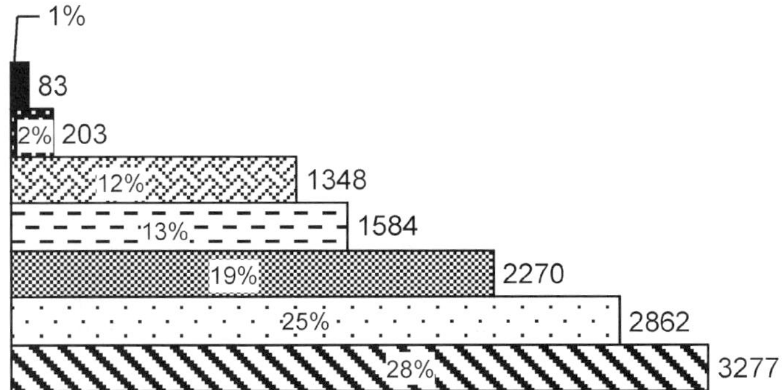

Figure 1.4 Plastics packaging by product (Western Europe, 1997). Total: 11 627 000 tonnes. Key: ■ others (extrusion, laminates); foams; thermoformed products; injection moulded products; sacks and bags; blow-moulded products; films. Units: × 1000 tonnes/year. Source: APME (data consolidated by product, resin and sector).

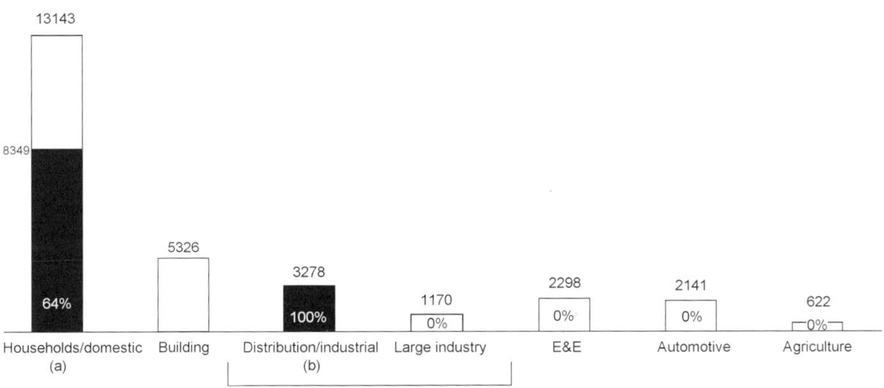

Figure 1.5 Plastics packaging by end-use sector (Western Europe, 1997). Total: 11 627 000 tonnes. (a) Food, detergents. (b) Packaging arising from supermarkets, shops, large industries, building and agricultural sites. Key: ■ packaging; □ non-packaging. Units: × 1000 tonnes/year. Source: APME.

wide array of moulding technologies, makes plastics an effective and exciting packaging medium, particularly for companies who are operating globally and are therefore seeking to harmonise packaging and reduce complexity.

There has been much rationalisation (joint ventures and mergers) in the polymer supply industry. At the time of writing, the two most recent are Dow/Union Carbide Corporation (UCC) and the BASF/Shell merging of the polyolefin business, operating as Targor, Montell and Elenac under the name Basell Polyolefin. More changes can be expected.

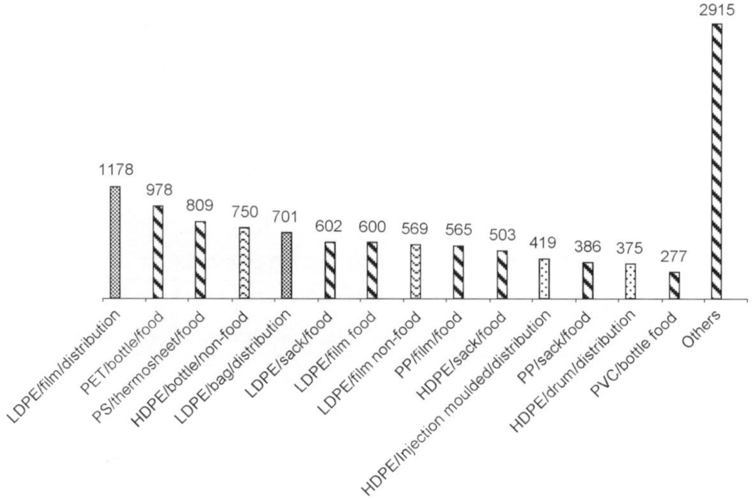

Figure 1.6 Main plastics packaging segments by end-use sector (Western Europe, 1997). Total: 11 627 000 tonnes. A segment is defined by the criteria resin × product × end-use sector. These main packaging segments (excluding 'Others') make up 75% of the total consumption of packaging. Key: ▩ distribution; ◣ food; ▨ non-food (mainly detergents/chemicals and household goods). Units: × 1000 tonnes/year. Source: APME.

Table 1.1 Major global polyethylene capacity (2000)

Company	Capacity (million tonnes)
Dow/UCC	8.5
Exxon/Mobil	6.4
Equistar	2.9
BASF/Shell	2.6
BP Amoco	2.4
Borealis/PCD	2.2
Formosa Plastics	1.7
Nova	1.6
Solvay	1.4
Phillips	1.3
Grand Polymer	1.2
DSM	1.2
Total Fina/Elf	1.1
ENI	0.8
Japan Polychem	0.7

The PET industry similarly operates on a global basis with most of the material being used for the manufacture of fibre and packaging. Major companies and production capacities are given in Table 1.3.

Plastics prices have traditionally been cyclical in nature, with the industry plagued by either under or overcapacity. It is, therefore, impossible to predict

Table 1.2 Major European polyethylene capacity (2000)

Company	Capacity (million tonnes)
Dow/UCC	2.6
BASF/Shell	2.4
Borealis/PCD	2.2
BP/Amoco	1.7
DSM	1.2
Total Fina/Elf	1.1
Exxon Mobil	0.9
ENI	0.8
Repsol	0.6
Solvay	0.5

Table 1.3 PET producers worldwide 1998–2000 ranked by capacity in 1998 ($\times 1000$ tonnes/annum)

Company	1998	2000
Eastman Chemical	1317	1452
Kosa	820	823
Shell	596	609
Du Pont	332	330
Nan Ya	268	361
Wellman	250	490
FET	190	280
Shin Kong	185	215
Rhodia-Ster	177	177
Kohap	171	171
Mitsui	170	180
Hualon	135	135
Tong Kook	129	129
Dow	120	280
SASA	100	100
Others	1948	2353
Total	**6893**	**8072**

Source: Du Pont Polyester

what will happen in the future but, at the time of writing, the prices of some of the common packaging polymers are given in Table 1.4.

The development of metallocene catalysts is having, and will continue to have, a major impact on the development of polyolefins for packaging. Not only are polyethylenes and polypropylenes with enhanced properties now available, along with some novel materials like syndiotactic polypropylene and cyclic polyolefins, but more can be expected in the future with the incorporation of other monomers into polyolefin chains.

Table 1.4 Polymer prices, November 1999

Material	Price £/tonne
HDPE: blow-moulding	630–680
film	630–660
LDPE	650–670
LLDPE	640–670
PET	550–606
PP: random copolymer	520–550
homopolymer	490–520
PS: general purpose	550–580
HIPS	580–610
PVC compound	530–550

1.2 Properties

The overwhelming success of plastics as a packaging medium can be attributed to the range of properties available and the existence of conversion technologies to maximise the properties of the pack while minimising the amount and therefore the cost of material. At the same time, plastics offer a range of marketing benefits. Most materials can be used in flexible, semi-rigid or rigid packs, further attesting to their versatility.

Plastic materials can be divided into two broad categories: structural materials which give form and the main strength to the pack, whether rigid or flexible; and materials which add a surface finish or a barrier, for example against the ingress of gases, such as oxygen, water or potential contaminants, or the egress of perfumes or flavour components. In many cases a single material like high density polyethylene (HDPE), linear low density polyethylene (LLDPE) and polyethylene terephthalate (PET) is all that is required to give all the necessary properties such as strength, barrier and product containment. For applications, such as food packaging, an additional barrier material may be required; for example, to keep out oxygen, which would accelerate product deterioration, small quantities of EVOH or nylon are often used.

Choice of material is also determined to a large degree by the conditions which the pack will endure, from manufacture to handling by the consumer. Storage of packs in freezers and distribution in winter require a pack material which can withstand low temperatures; a brittle material like polypropylene homopolymer would be a poor choice. For certain food products, filling is carried out at elevated temperatures, making material choice particularly critical.

For example, PET has many excellent properties, such as barrier, high gloss, high transparency and high purity, but has a low glass transition temperature that limits the upper temperature at which it can be filled. Manipulation of the crystallinity (heat set) and careful design of the container to avoid distortion have raised the filling temperature, but any developments which increase the

Table 1.5 Common polymer properties relevant to packaging applications

Property
Molecular weight (MW)
Molecular weight distribution (MWD)
Density
Intrinsic viscosity (IV)—particularly to PET
Melt flow index (MFI)—particularly to polyolefins
Tensile strength
Elongation at break
Vicat softening temperature
Transparency/opacity
Stress crack resistance
Barrier to water, gases and volatile components, e.g. perfumes and flavours

hot-fill temperature and improve the barrier to oxygen would further increase the use of this important material.

In flexible packaging, heat sealing, in some cases through the product or where some product contamination of the seal area is unavoidable during filling, is of fundamental importance. Materials with low sealing temperatures, a wide heat seal window and good seal strength are required. This has traditionally been the domain of low density polyethylene (or very or ultra low), ethylene vinyl acetate (EVA) copolymer or specialised ionomer materials like Du Pont's 'Surlyn'.

Metallocene developments have recently introduced a new class of polyolefins for this application—plastomers—with lower densities and melt temperatures than conventional LLDPE materials and more controlled sealing properties.

Some important material properties which will influence use of a particular polymer and which impinge on the properties of the final pack are given in Table 1.5.

1.3 Testing and specification

Plastics are complex materials, particularly in their interaction with products, and if a successful packaging development is not to become an expensive commercial disaster, rigorous testing of product and pack must be carried out. Many consumer products in commercial production are packed in very large numbers (hundreds of thousands or more). The price of failure is high, therefore careful attention to pack design, selection of materials and testing of the total pack with the fully formulated product is essential. The details of the pack then need to be enshrined in a specification which will serve to protect buyers, production in the end-use and the converter from expensive errors. Such a specification might include the following for a rigid container (many of the points will apply equally to flexible packs):

- dimensions, including critical dimensions in the neck region
- wall thickness
- top load/compression strength
- impact strength/drop test
- pigmentation type/level
- material: type (e.g. HDPE), approved grade(s), supplier(s) and
- layer configuration: e.g. mono, three (detail: inner/middle/outer)

Different types of product will require different levels of testing, but for aggressive liquids, moving from other types of rigid packaging, such as glass or metal to plastic, or a product moving from a rigid to a flexible pack, then substantial attention, including a rigorous test programme, is required to eliminate compatibility and environmental stress crack resistance (ESCR) problems.

1.4 Marketing considerations

Plastics packaging offers many marketing opportunities in the consumer goods industries in that it is frequently easier to revitalise a brand by changing the packaging than it is to change the product. The decision to use plastics in some extreme market sectors may be a difficult one. Many brands of spirits use plastic for airline and duty-free products because of weight savings, but such packaging might not have been considered optimum for general sales. Similar discussions are no doubt taking place in the beer sector on the merits of plastic *vs* glass, while developments to produce a PET beer bottle with the right barrier properties continue. The choice of material is quickly clarified by considering a few key marketing requirements:

- rigid or flexible
- single-serve (small) or multi-serve (large)
- transparent or opaque
- cost (material plus manufacture)
- timescale: machinery available
stock or custom pack
- handle/no handle
- other design features
- surface finish (glossy, etc.)
- product compatibility
- decoration options
- tamper evidence required
- product processing conditions (e.g. hot-fill) and
- shelf-life

At this point the only decision which frequently has to be made regarding the material is the most appropriate grade to meet compatibility requirements.

Factory and distribution requirements will also play a part in the decision-making process; for example, the pack may have to be compatible with an existing filling system or packing line. A move from a rigid load bearing pack to a flexible pack will have implications for secondary packaging and other elements in the distribution chain, such as whether to use cases or shrink-wrap collation.

Plastic containers in all their forms offer a wide range of marketing options—designs, colours, convenience aspects, tamper evidence, gloss, transparency and decorations. In-mould labelling can be used in both blow- and injection moulding. Soft touch containers and cans are available through the use of special grades of polypropylene or thermoplastic elastomers.

The range of effects which the materials can deliver are complemented by the available technologies and mould finishes. In short, plastics offer a wide range of properties which a marketer can use to differentiate a brand from that of the competition, if they are prepared to exploit what is available.

1.5 Rigid packaging

Rigid packaging covers items made by blow-moulding (in all its forms), thermoforming and injection moulding, with plastics being the material of choice for many applications where, for example, blow-moulded containers have replaced glass. Nowhere is this more obvious than in the phenomenal development of the PET beverage bottle, with continuing growth of around 15% per annum.

Thermoforming competes with injection moulding in the manufacture of thin walled containers for the food industry. Here fast cycling high melt flow grades of polypropylene are extensively used. Cycle time is critical in a highly competitive market and consistency of material properties is important. In other areas of injection moulding, co-injection and multi-centre moulding is used to exploit special material properties, creating new effects for the packaging developer.

The main targets for the PET producers are to improve oxygen barrier and increase the hot-fill temperature. The latter has been under development for a number of years, with PET/PEN blends being one option explored. Recent attention, however, has focused on the potentially lucrative beer market. Here the inability to fulfil the requirements of the brewing industry with regard to shelf-life has been a major stumbling block. Techniques such as co-injection to produce three to five layer structures with EVOH and nylon barrier materials, internal and external surface coating technique, such as Sidel's Actis Technology, and the use of PET/PEN blends are all under active consideration.

Oxygen absorbers are also being used, even in five layer structures such as the Miller Brewing containers which have been launched nationally. It is currently unclear which technologies will produce the most effective container, in terms of properties and cost, and meet the diverse recycling requirements of the US and Europe.

1.6 Flexible packaging

Many of the issues surrounding flexible packaging relate to use in food packaging and the need for consistently good or improved barrier properties, allied with good mechanical and optical properties. Flexible packaging has made an enormous contribution to improving hygiene and reducing wastage in food packaging in the developed world. Already, excellent material properties have been enhanced by modified and controlled atmosphere packaging, vacuum packaging and active packaging to increase shelf-life of sensitive products, retain flavours and deliver a more attractive product to the consumer.

Flexible bags (1–25 kg), shrink pallet and collation wrap, stretch wrap and other polyolefin films have benefited from developments in metallocene produced resins with enhanced properties and better strength.

Ease of opening (openability) has always been an issue with flexible packaging. Consider, for example, what has been achieved from the early experience with milk pouches. Much has been done to improve openability for packs from detergent liquids to fruit juices, with many pouches now having closing or resealable devices. Flexible pouches, as a result, are replacing rigid packs in some cases; for example, as refill packs and where material minimisation is seen as a benefit. This is seen as a growing market, particularly in the US. Many markets now have sports drinks packed in flexible pouches with rigid plastic necks and closures—a good example of a packaging format well suited to the market requirements, mainly safety and unbreakability.

1.7 Application areas

Plastic packaging is extensively used in the following sectors of the consumer products market:

- food,
- beverage,
- personal care,
- health and beauty, and
- household products.

Many of the key elements of plastics have already been touched on in this chapter and will be dealt with in more detail elsewhere. There are many properties of plastics packaging that are attractive to all the sectors, such as its lightness of weight, decoration options and the ability to use a rigid or flexible pack for many applications. Some of the salient features are summarised below.

Food packaging
Advantages

- Printing
- Range of standard and barrier options
- Transparent or opaque
- Wide range of properties, therefore shape and pack options available
- Lightweight
- Co-extrusion particularly important and
- Can be used in microwave

Disadvantages

- Even better barrier required for longer shelf-life and
- Higher filling temperature would further extend application range

Beverage packaging The main material here is PET, with the possibility of using polypropylene in some areas, particularly random copolymers.
Advantages

- Lightweight
- Unbreakable
- Can withstand high carbonation pressures (10–12 bar) and
- Very pure special low acetaldehyde grades for non-carbonated water

Disadvantages

- Creep
- CO_2 loss in smaller (330 ml) containers
- Oxygen barrier not good enough for beer and other oxygen-sensitive products and
- No integral handle

Personal care, health and beauty packaging
Advantages

- Unbreakable
- Attractive surface finish option and decoration
- Soft touch option
- Product resistance and
- Several material options

Disadvantages

- Barrier properties for perfume; absorption of perfume components into the walls of the container

Household products packaging
Advantages

- Wide range of materials
- Cheap
- Attractive surface and decoration options for branding
- Rigid or flexible
- Good chemical resistance and
- Good resistance to environmental stress cracking (with correct material choice)

Disadvantages

- None

1.8 Stock keeping unit (SKU) mix

When products are sold in more than one size, to preserve the brand family image, the primary pack should be made of the same material and preferably use the same technology for all the sizes. Refills may use differentiating pack types/technologies such as rigid primary, flexible (pouch) refill. Both options are available with plastic, the extreme range possibly being with HDPE, with blow-moulded containers ranging from 0.002–1000 l. With PET, the volume range is smaller but it more than encompasses the household products range.

Such flexibility also offers the possibility of single- or multi-serve packs, the former becoming increasingly important. In the US many ready meals now come in single-serve portion trays allowing the family a choice of menu to suit all tastes at the same mealtime.

Careful consideration needs to be given to the choice of plastic material, the moulding technology and the product life cycle in choosing the appropriate SKU mix. Too many variants will increase the cost of filling and distribution.

1.9 Costs

As mentioned earlier, polymer prices are cyclical in nature and the high price of a material at a point in time, particularly if coupled with more expensive conversion technology or filling line modifications, can effectively block the use of what may be the optimum plastic packaging material.

There are undoubtedly benefits in high volume use of a particular material or range of materials from one manufacturer, which is one of the reasons many large users of plastic packaging materials now operate globally. Even low volume end-users can benefit if a pack utilises a grade of material already consumed in large volume by the converter. Alternatively a move away from a high volume material, for technical reasons, to a single source or low tonnage purchase from distribution, can result in a substantial price premium.

Particular grades are specific to a particular supplier, but it is generally possible to find (with adequate testing) a range of materials with the right properties from different polymer manufacturers. Such a range of materials, therefore, offers not only commercial flexibility to the converter, but also security of supply in times of technical difficulty or shortage.

1.10 Support technologies

The success of packaging depends not only on materials and manufacturing equipment but also on a range of support technologies. Design is of fundamental importance to aesthetics, functionality and also to the ultimate proportion of the pack. No matter how good the material or manufacturing technologies, a poorly designed pack may fail before even reaching the marketplace. Design agencies therefore have an important role to play, but technical support is also required.

Finite element analysis can be used to determine the stresses in a pack at an early stage, before the design has left the computer. Features which are likely to cause technical problems with the pack, such as stress points, areas where extra material may be required to enhance properties, and implications for pack weight (important for cost and environmental reasons) can all be determined and the design refined to optimise the properties. Computer visualisation techniques can be used to view the design from all angles, or colour and shade the object, leading to further refinements and a faster decision-making process.

Stereolithography can produce the first model and can then be used in subsequent rapid prototyping developments to produce increasing numbers of samples for evaluation (10–1000 s). The precise mouldmaking technique employed will depend on the number of trial samples required.

Once the design has been fully validated, files can be transmitted digitally to the converter or mouldmaker for manufacture of commercial tooling. As speed to market is increasingly important in the packaging industry use of these technologies will increase.

1.11 Packaging without plastics—image, environmental issues and recycling

In spite of the overwhelming benefits of plastics as a packaging material, much has been written about the negative elements such as the fact that plastics come from oil, an unrenewable source (true), plastics do not biodegrade (true) and plastics cannot be recycled (false). Such statements grossly oversimplify the situation. Although plastics do indeed come from hydrocarbon sources, only 4% of oil is used in plastics production—most is used in energy generation.

The fact that most plastic packaging does not biodegrade should be seen as a benefit not a curse. The objective of packaging is to protect the product from time of manufacture to use by the consumer, with consumer safety being of paramount importance. Storage times can be long, transportation and humidity during storage can vary. It would be difficult indeed to control the rate of degradation in such circumstances without compromising the quality and safety, particularly of food and pharmaceutical products. There may be opportunities to use biodegradable plastics in niche areas, but even then for biodegradation to occur the right conditions must exist. These conditions are not present in landfill.

Plastics can be recycled provided the infrastructure is there to collect and separate them. Mechanical recycling is not the only (or best) option, unless a satisfactory end-use can be found for the recycled material. Chemical recycling and incineration with energy recovery needs to be considered, particularly for mixed plastics. A recycling route should not consume more energy or cause more pollution than the problem it is supposed to be solving. Life-cycle analysis (LCA) and related analysis of recycling and recovery routes are therefore imperative in identifying the most effective solution.

Packaging without plastics has serious implications for the environment. Without plastics, the following increase would occur in moving the same amount of product [1]:

- Energy +208%
- Weight +391%
- Volume +258%

1.12 Conclusion

A few commodity polymers such as polyolefins, polyethylene terephthalate, polystyrene and polyvinyl chloride, plus a number of specialised materials like EVOH, nylon and polycarbonate, can be used to develop a wider range of packaging, offering technical, marketing and environmental benefits.

Sophisticated technologies are available for high volume manufacture to exploit the material properties on a commercial scale. It is hard to imagine a world without the benefits of plastic packaging, which is why the market continues to grow. Product and consumer are protected, for example, by unbreakable soft drinks bottles.

Plastics are pure materials, not a threat to the environment and there are a number of recycling routes available. It is a matter of choosing the most appropriate and setting up the infrastructure to deal with it.

Advances in technology producing even better materials, and the demand for better properties, longer shelf-life and alternative packaging, will ensure that the use of plastics in packaging will continue to grow in the new millennium.

Reference

1. Gesellschaft für Verpackungsmarktforschung (GVM, Germany 1995).

2 Types of plastics materials, barrier properties and applications

D.W. Brooks

2.1 Introduction

Plastics are now the most widely used materials for packaging in terms of value—ahead of paper and board, metals, glass and other materials. The world packaging market is worth about \$400 million and is dominated by Western Europe and North America, each with about 25% share. In Western Europe, plastics account for 37% by value of materials used in packaging. The key polymers used for plastics packaging are polyethylene (PE), polypropylene (PP), polystyrene (PS) and polyethylene terephthalate (PET). These along with polyvinyl chloride (PVC) are regarded as commodity polymers. The use of PVC in consumer packaging has declined in recent years, especially in food and beverage packaging. PET is the major growth polymer and is currently growing at >10% per annum.

Rigid packaging accounts for the major share of packaging in consumer goods markets. In Europe, 5 million tonnes of polymers are converted into rigid packaging per year, with growth predicted to be 5% per year. The major polymers used for rigid packaging are, by volume: PE (35%), PET (20%), PS and EPS (19%), PP (15%) and PVC (11%).

Flexible packaging accounts for almost 2 million tonnes of polymers in Europe, with growth of 3% per year. The major polymers used for flexible packaging are, by volume: PE (52%) and biaxial oriented PP (BOPP) (27%), with cast PP, nylon, PVC and PET films having smaller demands.

The major markets for plastics in packaging in the consumer goods sectors are foods and beverages, which together account for about 65% of the total. The other consumer goods sectors are personal care, health and beauty, including cosmetics and household.

Blow-moulding accounts for 51% of the total volume of rigid packs, thermoforming 27%, injection moulding 18% and compression moulding 4%. The sectors accounting for the highest growth in rigid packs in consumer goods are: PET beverage bottles, PP beverage closures, pharmaceutical packaging, luxury beauty and cosmetics packs, dairy food products, and fast food packaging. The future growth sectors are likely to be beer (PET), mineral waters and hot-fill applications, convenience food packaging, pharmaceuticals, health and beauty packaging and modified atmosphere packaging.

Regulations and approvals are key elements for polymer producers, converters, packer-fillers and retailers, and it is important to be aware of requirements for compliance. This topic will not be discussed in detail, but everyone involved in the production, conversion and use of packaging—in particular, plastics for packaging—needs to know the importance of safety, environmental and health concerns.

One key issue which affects the use of plastics for packaging in the consumer goods sectors is that of food contact regulations. These exist in most countries, each with its own regulatory bodies and their way of regulating food contact applications. The major agencies are Food and Drugs Administration (FDA) in the US, with section CFR21, and the European Commission (EC) in Europe, with the Framework Directive 89/109/EC and the 'Plastics materials and articles intended to come into contact with food' Directive 90/128/EC. These agencies should be contacted for further information. Most packaging companies use food contact regulations as a standard for all packaging produced for the consumer goods sectors.

2.2 Commodity polymers

2.2.1 Polyester

Polyesters, of which group of compounds polyethylene terephthalate (PET) is the most common thermoplastic used in plastics packaging, are produced by reacting a dibasic acid with a diol (alcohol) to make an ester. PET is polymerised using the monomers terephthalic acid (TPA) and ethylene glycol (EG). PET has been used to make fibres for many years, but the high molecular weight PET needed to make film and bottle resin is relatively new (1978). Polyethylene naphthalate (PEN) has similar chemistry to PET, with naphthalic acid (NPA) replacing TPA to produce a semi-crystalline polyester with better gas barrier properties and higher mechanical properties than PET. PEN will be discussed in Section 2.3. Polybutylene terephthalate (PBT) has very similar chemistry to PET, replacing EG with butanediol to give a polymer used more in engineering applications, and less in packaging. Amorphous polyester glycol modified (PETG), another polyester used in packaging, is completely amorphous and has a distinct Tg (glass transition temperature) but no crystalline melting point. PETG is a copolyester produced using two diols, EG plus cyclohexenedimethanol (CHDM), which prevents crystallisation.

The total European market for PET resin in packaging is about 1.5 million tonnes. It is dominated in the consumer goods sectors by carbonated soft drinks bottles (44%) and bottled waters (33%), with smaller but growing markets in foods, other beverages (in particular juices, spirits), household, medical and healthcare, and toiletries and cosmetics. There is much research at the moment on improving the barrier of PET to make it suitable for oxygen-sensitive foods and beverages, such as beer and baby food. Polyesters can be processed by most of the common fabrication techniques into flexible (films) and rigid packaging.

2.2.1.1 Polyethylene terephthalate

Polyethylene terephthalate (PET) is a semi-crystalline (up to 60%) polymer. It has a high molecular weight and can be converted by injection, extrusion, and biaxially stretched blow-moulding into containers and bottles, via injection moulded preforms. The key criteria that determine the properties of PET are: molecular weight measured by intrinsic viscosity (IV), level of crystallinity, and Tg. PET is polymerised by a condensation process with the initial stage producing a low molecular weight polymer, called melt phase PET, which is amorphous and has a low IV. The PET resin is then solid-state polymerised using heat to increase the IV and crystallise the polymer. The IV determines the melt viscosity of the polymer, the processing conditions used for production of preforms and subsequently the stretching behaviour of the preform into the final bottle, and the physical properties, toughness and impact strength of the bottle. PET will not be fully crystalline, typically about 60% in the initial pellets (heat-induced) and about 20% in the final bottle (strain-induced), while injection moulded preforms will be completely amorphous. Stretching by orientation induces strain crystallisation and the formation of small crystals resulting in high clarity, increases the mechanical properties and reduces the gas permeability (improves barrier properties).

PET bottles produced via the stretch blow-moulding process are particularly suitable for packaging foods and beverages, and other products which require high clarity and gloss, a high barrier against gases, resistance to creep under pressure, and toughness and impact resistance. The major application is for the packaging of carbonated beverages. PET is a high melting polymer which readily absorbs moisture and must be dried prior to processing to avoid hydrolysis and degradation. It is important to process PET correctly and use conditions that do not cause degradation and the generation of acetaldehyde (AA), a gas which if present in the preform or bottle can migrate into products during storage, causing off-taste and flavour changes. This is particularly critical for non-carbonated mineral waters and cola drinks (see Table 2.1).

PET homopolymer is produced by reacting a single acid (TPA) with a single diol (EG). PET copolymers or copolyesters can be produced using a second diol (CHDM) or a second acid (IPA). PET homopolymers have a fast crystallising rate, a high melting point and a low natural stretch ratio for biaxial stretching. Copolyesters are slower crystallising, have a lower melting point, and higher natural stretch ratios. PET grades used for bottles usually have an IV at least 0.74 and more usually 0.8, and can be supplied natural or ready compounded, in green and brown.

Table 2.1 Polymers for blown bottles

	PET	PEN	PP random copolymer	PP homopolymer	HDPE
Transparency (%)	92	92	90	60	30
Oxygen barrier (cc.mm/m^2.day.atm)	1.5	0.3	90	60	50
Water barrier (g.mm/m^2.day 38°C/90% RH)	0.8	0.3	0.5	0.26	0.14
Stiffness (MPa)	2500	3000	1000	1600	1200
Impact strength notched[a] Izod (kJ/m^2)	No bk[b]	No bk[b]	7	3	16
Density (g/cm^3)	1.33	1.33	0.9	0.9	0.96

[a]notched: this refers to the modification of the test sample by notching to induce a more brittle like failure on impact testing.
[b]No bk: no break.

A major development of PET involves improving the gas barrier to make PET more suitable for packaging oxygen-sensitive and carbonated products. These developments will be discussed in Section 2.5.

2.2.1.2 *Amorphous polyester*

PETG is a copolyester, which is amorphous, does not crystallise, and has lower barrier properties than PET. Its advantage is that it can be extrusion blow-moulded into bottles and containers, and does not require the stretch blow process normally used for producing PET bottles. PET can be extrusion blow-moulded but the polymer needs to have very high molecular weight, and special equipment (screw, barrel) and process conditions are required, making it more expensive as a polymer and a process for making bottles.

2.2.1.3 *Heat set PET*

PET can be modified to improve its heat resistance to make it more suitable for packing hot-fill products or pasteurisable products. PET bottles can be heat set in a single process step using optimum stretching of the complete bottle with heated blow-moulds, or bottles can be heat set in a second step after blow-moulding, using heated blow-moulds. Cryogenic blow-moulding is a technique that uses liquid nitrogen during the stretching process with heated blow-moulds to produce a distinct transition of crystallinity between the outer surface and the inner surface of the bottle. This has been claimed to improve the barrier and mechanical performance of bottles.

2.2.2 *Polypropylene*

Polypropylene homopolymer is polymerised from propylene monomer, a gas derived from oil. PP is a semi-crystalline thermoplastic which has been extensively developed since it was first produced in 1951 by Phillips Petroleum Co. in the US. Today it is one of the most widely used polymers, being used worldwide in many packaging applications, for a complete range of consumer markets, such as the food, beverage, household, health and beauty and personal care sectors. PP has a wide property performance, making it a key choice for many packaging applications. PP can be made into flexible and rigid packs, film, containers, trays, bottles, tubes, caps and closures by all the major processes.

PP has a regular structure which is called isotactic, (iPP) with the methyl groups located on the same side of the chain. Other forms also exist. Syndiotactic PP (sPP), not available using Ziegler-Natta catalysis, is under development by Fina using a metallocene catalyst where the methyl groups are located alternately along the chain. The other common form is atactic PP (aPP) with the methyl groups randomly positioned along the chain. Previously a contaminant of isotactic PP, this can also be made 100% pure using metallocene catalysts.

Different types of PP can exist; the homopolymer is produced only from propylene monomer, while the copolymers have a second monomer copolymerised with propylene in the reactor. Random or Raco copolymer contains randomly distributed ethylene in the polymer chain. Table 2.2 shows PP polymer types.

The other common copolymer is the block or impact copolymer, correctly named 'Heco' or Hetrophasic copolymer. This is not strictly a copolymer, but a two-phase blend of PP and rubber phases (normally ethylene propylene rubber), the rubber forming blocks along the propylene chain. PP grades are usually specified according to the melt flow for processing by a particular technique. Low melt flow grades have a high molecular weight, and are used in extrusion blow-moulding processes to produce containers, bottles and jars, and also in sheet extrusion grades for thermoforming into trays and pots. High flow grades have a lower molecular weight and are used for injection moulding and compression moulding of caps and closures, cups, containers and pots. Very high flow grades are used for thin-walled injection moulded containers, cups and pots. Table 2.3 shows the polymer properties of polypropylene.

Table 2.2 PP polymer types

Homopolymer (HOMO)	PPPPPPPPPPPPPP
Impact or block or hetophasic copolymer (HECO)	PPPPP**RRRR**PPPP
Random copolymer (RACO)	PP**E**PP**E**PPPPP**E**PP

R is ethylene propylene rubber. E is ethylene.

Table 2.3 PP polymer properties

	PP homopolymer	PP impact copolymer	PP high impact copolymer	PP random copolymer
Density (g/cm^3)	0.905	0.905	0.905	0.90
Stiffness (MPa)	1600	1400	1200	1000
Impact strength notched Izod (kJ/m^2)	3	10	30	7

2.2.2.1 *Homopolymer*

PP homopolymer has high crystallinity and high stiffness but low impact strength. It is easily processed using all common process techniques and has good temperature resistance. It can be used for sterilised packed products and is a good moisture barrier, but is a poor barrier to gases like oxygen and carbon dioxide. It is not suitable, alone, for long shelf-life oxygen-sensitive foods or beverages. PP is widely used for caps and closures, including closures for pressurised bottles (glass, PET) used for carbonated drinks. Because it has a natural hinge performance, it can be used for closures and packs with a hinge mechanism. PP competes with HDPE for closures intended for beverage applications which require high torque resistance on pressurised bottles for carbonated drinks. Table 2.4 shows polymers used for caps and closures.

Table 2.4 Polymers for caps and closures

	PP homopolymer	PP impact copolymer	HDPE
Stiffness (MPa)	1600	1400	1200
Impact strength notched Izod (kJ/m^2)	3	10	16
Stress crack resistance	Excellent	Excellent	Good–fair
Living hinge	Excellent	Excellent	No
Temperature range (°C)	20–100	5–100	−20–60

The most important recent developments have been grades with increased stiffness (rigidity) and increased impact strength. Borealis has recently introduced ‘controlled crystalline PP’ (CCPP) grades in both homopolymer and copolymer. These polymers are nucleated in the polymerisation reactor stage and not by using nucleating additives.

Grades now exist with stiffness approaching that for ABS (2300 MPa) so that down-gauging and lower wall thickness will bring all round benefits to converters and end-users of films, bottles and containers, with the potential for improving the gloss and transparency for packaging such things as cosmetics, personal care and foods. Montell have developed their ‘Adstif’ grades with very high stiffness.

2.2.2.2 *Impact copolymer*

PP block or impact copolymer is a high crystalline two-phase system of PP with low levels of ethylene propylene rubber (EPR) which improve the impact performance of PP. Impact copolymer can have high stiffness with high impact in particular at low temperatures (below 0°C). It is easily processed using most techniques, has good temperature resistance and offers a good moisture barrier. It is not a good gas barrier, so it is not suitable for packaging oxygen-sensitive products. Impact PP is opaque; the two-phase system with a rubber phase scattering light and therefore reducing clarity. It is now possible to produce the rubber *in situ*, directly polymerised in the reactor so that a finely dispersed EPR is formed directly with the PP. This results in a high impact copolymer with a notched impact strength ($>20\,kJ/m^2$) while retaining high stiffness (1600 MPa).

Impact copolymer PP is widely used for caps and closures, in food packaging (caps for food containers and jars) and for beverage closures (in particular for pressurised carbonated drinks which are packed in glass and PET bottles). Because PP is stiffer and harder than HDPE and does not give an elastic seal, it is necessary to incorporate a soft liner (usually EVA) to make the seal and provide a high removal torque. PP is also widely used on various packs including those for food, household, personal care and cosmetics where the cap or closure requires a hinge mechanism (see Table 2.4).

2.2.2.3 Random copolymer

PP random copolymer is made from propylene with the addition of low levels of ethylene, or other monomers. It has lower crystallinity than homopolymer and impact PP polymers. It usually has a lower stiffness with medium impact strength which is better than homopolymer. It is easy to process, has a lower temperature resistance, is a good moisture barrier, but a poor gas barrier. Excellent clarity is possible when nucleated with special clarifying agents to produce small crystallites in the structure. Nucleating additives which are used as clarifiers include sorbitols and derivative materials.

Random PP copolymers, both nucleated and non-nucleated, are finding applications where good transparency and gloss are required in bottles and containers for packaging foods, personal care and cosmetics to compete with PET and other clear polymers (see Table 2.1).

2.2.2.4 Soft flexible copolymer

The demand for flexible PP grades has increased dramatically in recent years. There are a number of different types which include modified random copolymers, atatic PP and new super flexible PPs.

RAHECO or random heterophasic propylene–ethylene copolymers have a matrix which is a random PP copolymer and the second phase an EPR rubber, with both the matrix and the rubber phase having varying degrees of flexibility, and a finer rubber distribution. The modulus range of these polymers extends from about 600 to 150 MPa. The high melting point of PP is retained, and so the heat resistance is maintained for packaging hot products and sterilising foods. These polymers can be injection moulded into thin-walled containers, with improved clarity, for medical and food packaging, or for packs which require an integral hinge. They can be blow-moulded into containers, bottles, tubes, and used as a ‘soft touch’ layer on bottles, containers and tubes to give a tactile effect for personal care and cosmetics packs. Montell has developed a range called ‘Adflex’.

Syndiotactic PP (sPP) is a direct result of the capability of new single-site catalysts, like metallocenes, to tailor the structure of polymers. The methyl groups alternate in sPP, rather than occur on the same side of the chain, as in iPP, resulting in a dramatic change in the properties. Compared to conventional Ziegler-Natta catalysed iPP homopolymer, sPP has a much lower melting point (30°C lower), is more flexible (500 MPa) and has good clarity and very low haze. sPP also retains its properties after exposure to gamma radiation, and has low heat seal temperatures compared to random PP copolymer. sPP is expected to be complementary to current PP grades, and to be used in completely new applications, including packaging, alone or in blends with PP. Fina is developing sPP polymers, which have yet to be used commercially.

2.2.3 Polyethylene

Polyethylene (PE) is polymerised from ethylene monomer. High density polyethylene (HDPE) is sometimes referred to as linear polyethylene, because the polymer chain contains only a few branches, allowing the chains to pack tightly together and giving high crystallinity (75%). Linear polyethylenes are made by low pressure polymerisation of ethylene. Short chain branching is a result of the introduction of a second monomer, an α-olefin (1-butene, 1-hexene). Medium density PE (MDPE) has a slightly lower density than HDPE, and linear low density PE (LLDPE) contains more comonomer and is more branched, and hence has a lower density. Higher levels of comonomer can be introduced to produce very low density PE (VLDPE). Branched or low density PE (LDPE) is made by a high pressure polymerisation; the homo- and copolymers form different length branched chains. LDPE is very highly branched, and the chains do not pack tightly together, giving lower crystallinity (45%) than HDPE, and thus much better transparency.

Most of the key properties of PE polymers (Table 2.5) are linked to the density or crystallinity of the polymer, so that an increase in density will result in an increase in stiffness, hardness, strength, heat resistance and chemical resistance. A decrease in density will increase flexibility, stress crack resistance (ESCR), creep, permeability to gases and moisture, and impact resistance. PE is the most widely used polymer for all types of packaging, particularly in the consumer goods sectors.

Table 2.5 PE polymer properties

	HDPE	LDPE	LLDPE
Density (g/cm^3)	0.958	0.920	0.926
Stiffness (MPa)	1200	200	400
Impact strength notched Izod (kJ/m^2)	16	40	40

2.2.3.1 Linear high density

Linear PE (HDPE) has a high density (about 0.96 g/cm^3), is highly crystalline and has high stiffness. The homopolymer is produced from ethylene monomer alone, while copolymers incorporate a second comonomer, such as 1-propene, 1-butene or 1-hexene. The molecular weight (MW) is a measure of the polymer chain's size, and determines the ease of processing and the mechanical properties of the finished component. A high MW implies high viscosity for the polymer melt when processed, and gives mechanical properties.

HDPE grades are obtained from a range of different catalyst systems, and will result in polymers with different structures, and different MW distributions (MWD). MWD is linked to both ease of processing and properties of finished components. Narrow MWD grades are needed for injection moulding, to ensure components are free from internal stresses and distortion, to prevent ESCR. Broader MWD grades are better suited to blow-moulding and film production (see Table 2.1).

HDPE is used in all the consumer goods sectors: particularly for bottles, containers, caps and closures in rigid packaging and films, bags, pouches in flexible packaging, for a wide range of products. It is also used for the production of beverage closures for carbonated drinks which require resistance to high pressures and have good torque resistance. It is particularly important to ensure the correct selcction of HDPE grades used for packaging personal care and cosmetic products, such as oils, detergents, soaps, emulsifiers, alcohols, and acids which may cause stress cracking (ESCR) (see Table 2.4).

In general, ESCR improves with a decrease in density or crystallinity, and increase in MW. HDPE is used in the household sector for packing chemicals, motor oils and cleaning products.

2.2.3.2 Branched low density

Branched PE (LDPE) has a lower density (about $0.92\,g/cm^3$) than HDPE and a maximum crystallinity of about 45%. LDPE is highly branched which leads to a low melting point (about 110°C) and flexibility. LDPE can be made into flexible and semi-rigid packaging, films, bottles, tubes, containers, caps and closures, bags and pouches by extrusion, injection, blown film and extrusion blow-moulding. Small bottles, tubes and containers are produced for health and beauty packaging, personal care packaging and cosmetics. The properties specified are flexibility and gloss with some degree of transparency. LDPE is widely used for pharmaceutical packaging because of its flexibility and clarity; examples are eye drop bottles and caps, nasal spray bottles and caps.

2.2.3.3 Linear low density

Linear low density PE (LLDPE) is a copolymer of ethylene with a second monomer, which is an α-olefin, either 1-butene, 1-hexene or 1-octene. LLDPE has short branches of the comonomer attached to the PE main chain. It has a density similar to LDPE (about $0.92\,g/cm^3$) and the introduction of the comonomer reduces the crystallinity. Very low densities below $0.9\,g/cm^3$ are possible with high comonomer contents when single site catalyst systems like metallocenes are used. Polymers produced using metallocene catalysts will be discussed in more detail in Section 2.5.

The properties of LLDPE copolymers are linked to both density and comonomer content. An increase in density gives higher crystallinity, increased stiffness, hardness, strength, impact strength, heat resistance, chemical resistance, better creep resistance, and lower permeability to gases and moisture. Increasing the comonomer content generally results in a decrease in these properties.

2.2.4 *Polystyrene*

Polystyrene (PS) is an amorphous polymer made from styrene monomer. It is a very transparent polymer, sometimes referred to as general purpose or 'crystal' polystyrene. PS has good mechanical strength but very poor impact properties, and so it is modified with rubber (butadiene rubber) to produce impact modified PS or high impact PS (HIPS). HIPS is produced by polymerising styrene in the presence of rubber, forming PS as the matrix phase and finely distributed rubber particles as the elastomeric phase within the PS. A wide range of rubber modification is possible with HIPS, giving low, medium and high impact grades. Another type of PS polymer is the styrene butadiene block copolymer (SBS). This has good transparency like PS, but also good impact performance like HIPS. SBS copolymer is made by copolymerising styrene and butadiene to form chains of alternating blocks of PS and butadiene rubber. A two-phase structure is formed but because the phases are very small, light can still pass through the polymer with little scattering and so the polymer remains clear. PS polymers in general do not have good gas barrier or moisture barrier properties, so are not suitable for oxygen-sensitive products. PS polymers are widely used in the food packaging sector for short shelf-life products or less oxygen-sensitive products. Table 2.6 shows the polymer properties of PS.

Table 2.6 PS polymer properties

	GPPS	HIPS	SBS copolymer
Density (g/cm^3)	1.05	1.04	1.01
Stiffness (MPa)	3500	2700	1500
Impact strength notched Izod (kJ/m^2)	<1	7	5

2.2.4.1 *General purpose polystyrene*

General purpose or crystal PS (GPPS) is a very clear polymer. It is hard, stiff, has good dimensional stability, but poor impact performance, and is very easy to process by most technologies including injection, extrusion, and in extruded sheet form by thermoforming. A wide range of grades is possible from high MW with high strength, high heat resistance, and easy flow with good strength, and very easy flow (low MW) for high speed injection of thin-walled packaging containers. Specially developed grades with low environmental stress crack resistance (ESCR) offer better performance under exposure to stress agents such as fats and oils in food packaging. Blends of PS with PE have improved moisture barrier properties compared to PS for the manufacture of lids and containers for food packaging. PS can be easily printed on and coated, making it especially useful for packaging.

The range of packaging made from PS is extensive. The major products in the consumer goods food sector are dairy product packaging, expanded PS meat trays, dessert, yoghurt and ice-cream packaging and drinking cups.

2.2.4.2 High impact polystyrene

HIPS is opaque, has good impact resistance, can vary in stiffness and offers a range of finishes from high gloss to satin matt. The two-phase system with rubber particles prevents any clarity but otherwise the properties are very similar to PS. A wide range of grades can be made by varying the rubber content to give low, medium or high impact performance. Varying the MW gives different flow grades, including very easy flow with various degrees of toughness for high speed injection of thin-walled food packaging. Grades with high heat resistance, high ESCR and high gloss with enhanced scratch resistance are used for caps, and containers for personal care products. HIPS is the ideal polymer for packaging dairy products in the food sector, with its efficient processing in extrusion into sheet and thermoforming. It is cost effective, has good toughness and rigidity, good low temperature impact strength and heat resistance for hot filling. PP has challenged HIPS for this market for many years, with increasing amounts of PP replacing polystyrene for yellow fats containers.

2.2.4.3 SBS copolymer

Styrene butadiene block copolymer (SBS) has good clarity and is tough with good impact strength. It is easily moulded and can be converted by injection, extrusion and, via sheet, is used in thermoforming. Typical applications in packaging in the consumer goods sector include: caps and closures, snap lids, jars, containers with integral hinges and cups. In the food sector extruded sheet can be converted into snap packs, salad packs, meat packs and lunch boxes. SBS can be blended with PS in biaxial oriented transparent co-extruded multilayer barrier films for food packaging. SBS can be converted into blown bottles and blown film using high melt strength grades. It is only very slightly less transparent than PS but has similar gloss which is an improvement on any other PS copolymer. PS copolymers are not suitable for packaging oxygen-sensitive food products. The barrier to gases and moisture is slightly better than that of PS but it is not considered a high barrier polymer. SBS like other styrene-based polymers does not have good resistance to some oils, but generally it is superior to PS and sometimes ABS. SBS copolymers are produced by BASF under the trade name of ‘Styrolux’ and by Phillips as ‘K Resin’.

2.2.5 Polyvinyl chloride

Polyvinyl chloride (PVC) has been used in packaging for many years. It is an amorphous polymer with only a small level of crystallinity. Although high volumes are used globally in packaging—over 3 million tonnes per year, with about 70% going into rigid packaging for bottles, blisters and clamshells—the use of PVC in packaging is generally declining. This decline has been the result of environmental pressure, adverse cost relative to competing PET and PP, and with advances in the materials and processing development in extrusion blow, sheet extrusion and thermoforming equipment.

Vinyl chloride homopolymers are manufactured by emulsion, suspension, or mass/bulk polymerisation. Copolymers are produced by emulsion and suspension, and also by graft polymerisation. PVC is unlike other thermoplastics in that it is not ready to use for moulding. The polymer must be prepared by mixing or blending with other materials and additives, necessary for processing and for a particular application. PVC is usually supplied as a dry powder blend with all the additives included, or as a pelleted compound again with all the additives included in the material. PVC can be supplied for rigid application without plasticiser (PVC-U), or with plasticiser (PVC-P) for flexible applications. The molecular weight of the PVC molecule determines the end-use properties of the material; the higher the MW the better the mechanical properties, but the more difficult the polymer is to process. PVC-U is mostly used for packaging applications and can be processed by extrusion into sheet and film, calendered into sheet, extrusion blow-moulded and extrusion stretch blow-moulded into bottles and containers, and injection moulded. The impact strength of PVC-U can be improved by the addition of impact modifiers (acrylate elastomers, MABS, MBS or EVA copolymers), producing good clarity bottles suitable for packaging applications for food, beverages and cosmetics. PVC-U can be extruded into blown film and sheet, and then thermoformed into pots, cups, trays and blisters for food packs. PVC has excellent chemical resistance and resistance to fats, oils and alcohols.

PVC-P plastisols are specially formulated and foamed for use as a seal material in metal closures, used for a wide range of food products packed in glass. This is a major use for PVC-P accounting for many billions of closures. An alternative, viable non-PVC seal material has so far not been identified.

PVC remains a widely used polymer for pharmaceutical and food confectionery packaging, such as blisters formed from calendered and extruded sheet. With the move away from PVC to PET for bottles in Europe, other continents like Asia and Africa are seeing growth in the conversion of PVC for bottles. PVC once dominated the bottle market in Europe for food and beverages, such as non-carbonated water, juices, wine and edible oils. With the lower density and, more recently, the lower price of PET, it was only a matter of time before PVC lost to these markets. One large market for bottles until recently was that of non-carbonated mineral waters, but bottlers moved away from glass to PET for sparkling waters, and the switch to PET for non-carbonated water soon followed.

In the personal care and cosmetics sectors there is still some growth, especially for custom design, short runs and smaller, complex shapes (see Table 2.1).

2.3 Barrier polymers

2.3.1 Ethylene vinyl alcohol copolymer

Ethylene vinyl alcohol copolymers (EVOH) have been manufactured and commercially available since about 1972. Currently there are two major producers, both Japanese companies: Kuraray produce 'EVAL' resins, and Nippon Gohsei produce 'Soarnol' resins; both are used in packaging films and for barrier layers in multilayer structures. EVOH is a random copolymer of ethylene and vinyl alcohol. It is crystalline and can be processed by extrusion and injection technologies to produce an excellent gas barrier material, suitable for food, beverage, medical, cosmetics and household packaging. The special properties are achieved when the correct copolymerisation ratio of ethylene to vinyl alcohol is used.

In addition to its unique gas barrier properties, EVOH also has excellent oil and organic solvent resistance, so it is used in packaging for oily foods, edible oils, mineral oils and organic solvents. It offers a good barrier to aroma and flavour and is therefore effective at retaining fragrances, preserving the aroma and flavour of the contents within packaging, and preventing undesirable odours penetrating the pack from the exterior. EVOH copolymers have a high gloss and low haze, which can produce good clarity in films and packs.

EVOH can be processed on conventional extrusion and injection equipment into blow or cast film, sheet, extrusion blow and co-extrusion blow-moulding, extrusion coating, injection and co-injection moulding. The copolymers can be processed with a range of other polymers including PE, PP, nylons, PS and PET.

EVOH copolymer grades are characterised by their ethylene content, melt flow and gas barrier properties. Standard grades usually have an ethylene content between 32 and 44 mol%; the high ethylene content copolymers have a lower density, lower melting and glass transition temperature (Tg) and lower (worse) gas barrier. EVOH has excellent gas barrier properties which are superior to any other thermoplastic polymer used for packaging applications, in particular when dry. The gas barrier properties will vary with ethylene content, but are also affected by moisture. EVOH has –OH groups in its structure which makes the copolymer hygroscopic and therefore it absorbs moisture. The gas barrier properties are adversely affected by absorbed moisture, so packaging applications involving high humidities either during processing (sterilisation, retorting, pasteurisation, hot filling) or in use need to be closely monitored.

Table 2.7 High oxygen barrier polymers

	Oxygen permeability (cc.mm/m^2.day.atm)	
	0% relative humidity	100% relative humidity
LCP	0.003	0.003
PVDC	0.004	0.004
EVOH	0.003	0.25
N-MXD6	0.09	0.15
PA6I/T	0.5	0.2
PEN	0.3	0.3
PET	1.5	1.5

However, even under high humidity conditions, EVOH still has excellent gas barrier properties, which will recover with time. Co-extrusion and co-injection of an EVOH layer between layers of high moisture barrier polymers like PE and PP will reduce the loss of barrier properties. This is now widely used in packaging applications for foods and beverages. Because EVOH is hygroscopic, its moisture barrier performance is not as good as some other polymers used in packaging. It is recommended that high ethylene content copolymers are used for applications where a moisture barrier is required (see Table 2.7).

Examples of flexible packs in which EVOH is used include co-extruded films and bag-in-box packs for foods, wine and milk powder, which include multilayer structures with LDPE, HDPE, LLDPE and EVA polymers. Rigid packs include bottles co-extrusion blow-moulded with PP, HDPE, LDPE and PET for a wide range of foods like ketchup, mayonnaise, dressings, baby foods, beverages, carbonated soft drinks (CSD), juices and more recently for beer. Cups and trays are co-extruded from sheet and thermoformed with PE and PP again for a wide range of foods, and also tubes for cosmetics and health products.

2.3.2 Polyvinylidene chloride

Polyvinylidene chloride (PVDC) is a copolymer of vinylidene chloride (VDCM) with other monomers including vinyl chloride (VCM), acrylic esters and carboxyl containing monomers. The two major producers of PVDC copolymers are Dow with 'Saran' grades and Solvay with 'Ixan'. The content of VDCM in the polymer is usually between about 70–90 mol%. PVDC is a semi-crystalline barrier copolymer and the volume used in packaging is low compared to commodity polymers like PVC. The total volume used in Europe represents less than 0.5% of the total polymers used for packaging. PVDC has excellent gas barrier and moisture barrier properties, and it can be processed by extrusion techniques, usually as a barrier layer in multilayer films and rigid packs. PVDC is mostly used in flexible packs; it is currently used less for rigid packs like bottles, trays and cups, having been replaced by other barrier polymers such as EVOH and barrier nylons. The copolymer can also be used as a water-based

Table 2.8 High moisture barrier polymers

	Moisture permeability (g.mm/m^2.day 38°C/90% relative humidity)
LCP	0.004
PVDC	0.01–0.1
PCTFE	0.015
COC	0.1
HDPE	0.14
PP	0.26
PEN	0.3
PET	0.8

dispersion or solvent-based coating for application to the external surface of containers like PET bottles to improve the gas and moisture barrier. The consumer goods markets in which PVDC is used include foods, some beverages and pharmaceuticals. The material can be co-extruded with PVC, PE, PP and PS. PVDC coated PVC is used particularly in the pharmaceutical industry where a good moisture barrier is required.

PVDC does have benefits over other competing barrier polymers: it has excellent gas barrier and moisture barrier properties, as well as being a good chemical barrier against aroma and flavours, and it is resistant to oils and fats. It does, however, have particular disadvantages in that it is more difficult to process, requires special equipment for extrusion and cannot be processed by injection moulding. Like PVC, it is prone to environmental criticism. These issues have restricted its growth as a polymer for packaging at the expense of other barrier polymers (see Tables 2.7 and 2.8).

2.3.3 Polyesters

PET alone does not provide enough of a barrier for highly oxygen-sensitive or moisture-sensitive products. Previous methods of improving the barrier properties involved coating the outer surface of a PET bottle, film or sheet with a high barrier polymer like ethylene vinyl alcohol copolymer (EVOH) or polyvinylidene chloride (PVDC). In the early 1980s Metal Box Co. (now Carnaud Metalbox, part of Crown Cork & Seal) developed a spray-dip coating process for producing a thin coat of PVDC on the outer surface of PET bottles. This gives a modest improvement in barrier properties and extends the shelf-life of beer in large bottles by a few weeks. The alternative, which followed some years later, was the development of high barrier polyesters which could either replace, blend or be co-injected with PET to give a barrier multilayer structure.

Eastman Kodak Chemicals developed the first PEN polymer called '5XO' which provided an oxygen barrier that was five times better than that of PET. At about the same time, Goodyear produced their PEN polymer 'HB' which had similar properties. The high price and limited monomer availability prevented full commercialisation at this time.

The options today are even more varied and barrier improvements can be achieved in a number of ways. Multilayer preforms can now be produced in multicavity injection tools or by over-moulding with a barrier polymer. Barrier polymers can be blended with PET. Coatings can be applied to PET bottles both internally and externally, using both inorganic and organic materials. The coatings can be applied by spray-dip techniques or by deposition via plasma technology. Polymer and non-polymer additives, fillers and scavengers can be added, blended with PET or incorporated as a barrier material as a separate layer. Barrier developments are discussed further in Section 2.5.

Table 2.9 Oxygen barrier of commodity polymers

	Oxygen permeability (cc.mm/m^2.day.atm)
LDPE	200
PS	150
PP	70
HDPE	60
PVC-U	4
PET	1.5

2.3.3.1 Polyethylene naphthalate

Polyethylene naphthalate (PEN) is a high barrier polyester with a better gas barrier, better UV barrier and higher stiffness than PET, but at a higher cost. PEN is polymerised using 2,6-dimethyl naphthalenedicarboxylate (NDC) as the key monomer. PEN is being developed by Eastman, Shell (ex Goodyear) and Kosa (ex Hoechst). It is a semi-crystalline polymer similar in structure to PET and has a high molecular weight, so it can be converted by injection, extrusion, and biaxially stretch blow-moulding into containers and bottles via injection moulded preforms. PEN is a higher melting polymer than PET, otherwise the ease of processing is similar to that of PET. After initial objections to the use of PEN for bottles that could end up in the recycle stream mixed with PET, technology was developed to separate the two polyesters, so that now PEN is accepted by the FDA for food contact applications. PEN is still an expensive polymer for packaging, mainly because of the high price for the monomer, and the consequent limited capacity, which will only increase with increased demand. These limitations have led to the development of copolymers and blends of PET/PEN that offer improved barrier compared to PET at a lower cost. Low PEN content copolymer with typically 10% PEN offers the potential of making a pasteurisable beer bottle, while high PEN content copolymer with 90% PEN would be suitable for hot filling, and is similar in cost to PEN homopolymer. PEN is being targeted at food and beverage applications where PET can not meet the oxygen gas barrier requirements, heat stability for hot filling, retorting or pasteurisation. PEN is being test marketed for beers and returnable packs. PEN/PET blends have so far not been commercially significant (see Table 2.7).

2.3.4 Nylon-MXD6

The key polyamide or nylon used for barrier packaging is a polymer called 'Nylon-MXD6' produced by Mitsubishi Gas Chemical (MGC). It is a semi-crystalline polymer, with high stiffness, good transparency and a high gas barrier. This barrier nylon is produced by polycondensation of *meta*-xylylene diamine (MXDA) with adipic acid. It is a polymer with some particular properties which distinguish it from other nylons like nylon 6 and 66. It has higher tensile strength and modulus, higher Tg, lower water absorption and excellent gas barrier properties. N-MXD6 exhibits excellent gas barrier properties compared to other nylons and many other thermoplastics, and is comparable with EVOH and PVDC copolymers. It has a wide processing window so it can be processed by injection, co-injection, extrusion and co-extrusion techniques, in combination with other polymers like PET, PP and PE to produce multilayer structures such as film for flexible packaging and bottles for rigid packaging applications. Like most other nylons N-MXD6 is hygroscopic and absorbs moisture, and its gas barrier properties deteriorate under high humidity conditions. It is more moisture-sensitive than PVDC, but less sensitive than EVOH (see Table 2.7).

N-MXD6 polymer grades are characterised by melt flow, for processing and conversion into film, sheet or containers. Its high melt temperature makes it particularly suitable for processing in multilayer structures with PET, other nylons and PP. Like PET N-MXD6 can be biaxially stretched and thermoformed to produce oriented high barrier multilayer flexible films, and injection moulded with PET into preforms for stretch blown bottles, for packaging oxygen-sensitive foods and beverages. It can also be blended with other polymers, in particular other nylons and PET, to produce intermediate barrier containers. N-MXD6 is being used for the development of high gas barrier, multilayer, PET bottles for beers and other beverages (see Table 2.10).

Table 2.10 High oxygen barrier films

Film (20 μm at 20°C)	Oxygen permeability (cc/m^2.day.atm 0% RH)	Oxygen permeability (cc/m^2.day.atm 85% RH)
EVOH 32% ethylene	0.13	1.5
EVOH 44% ethylene	1.3	3.3
N-MXD6 oriented	3	3.6
Nylon6/PVDC coated	7	7
PP/PVDC coated	11	11
OPA6	25	65
OPET	60	60
OPP	2000	2000

2.3.5 Other barrier polymers

2.3.5.1 Acrylonitrile copolymer

Acrylonitrile copolymers have been developed, commercialised and have disappeared over the past 20 years. Today only one material is commercially used in packaging and that is 'Barex', an acrylonitrile–methyl acrylate copolymer grafted onto nitrile rubber, produced by BP Amoco. Barex is an amorphous high nitrile content copolymer. Polyacrylonitrile (PAN) is a brittle, amorphous, transparent polymer with excellent gas barrier properties, and chemical resistance, but in its pure form it is difficult to process. It has not been produced commercially for packaging applications. Copolymerising AN with other comonomers, such as methyl acrylate, onto a nitrile rubber backbone provides impact strength and produces a polymer which retains most of the key properties of AN, but is easier to process. The refractive index of the nitrile rubber is matched to the copolymer matrix to ensure good clarity.

'Barex' has good gas barrier properties, high stiffness, good impact strength and good clarity, although it has a brownish tint. It does not have high temperature resistance or high moisture barrier properties. 'Barex' is suitable for processing by the following techniques: injection, injection blow-moulding and extrusion, extrusion blow, extrusion stretch blow, injection stretch blow, sheet extrusion and thermoforming. It can be moulded into monolayer containers and bottles. It has a wide processing window and high melt strength for deep-draw thermoforming and bottle production. It has very good resistance to oils, flavours, odours, essential oils, chemicals and solvents. It can be used in a wide range of packaging applications including foods, beverages, cosmetics, personal care, pharmaceuticals and household products. The largest market for 'Barex' is thermoformed packaging for food, medical and chemical products. 'Barex' has been selected for packaging long shelf-life fruit juices in bottles in preference to multilayer bottles containing a barrier layer, because it offers better gas barrier properties, flavour retention and clarity. Other packaging applications include food products (oils, sauces, spices and delicately flavoured foods), cosmetics and personal care products (nail polish, perfumes, bath oils, toothpaste) and households products (pest control products and air fresheners).

2.3.5.2 Polyketone copolymer

BP Amoco has also developed another barrier polymer called 'Ketonex', a polyketone copolymer produced from polyolefin polymers and carbon monoxide. The term copolymer relates to alternating ethylene:carbon monoxide polymers, and terpolymer relates to olefin:carbon monoxide polymers which include ethylene and a second olefin monomer, like propylene. Polyketones have a unique combination of high temperature resistance, chemical resistance, gas barrier properties and low wear properties. 'Ketonex' is a semi-crystalline polymer which has excellent gas barrier properties, similar to EVOH, but it is potentially cheaper because it is produced from low cost monomers. A number of packaging applications have been identified. 'Ketonex' has been successfully processed by injection and extrusion into film, sheet and bottles as mono- and co-extrusion multilayer structures with PE, PP and PVC. Some packaging developments have been undertaken with selected partners, but to date no commercial applications have been identified.

BP Amoco has now announced (October 1999) that it will divest its aliphatic polyketone technology, and is looking for a strategic fit with another company. 'Ketonex' is a very interesting polymer for packaging, but requires further development before it becomes commercially viable.

2.4 Technical polymers

2.4.1 Olefin polymers

2.4.1.1 Ethylene vinylacetate copolymer

Ethylene vinylacetate copolymers (EVA) are olefin polymers with the addition of vinylacetate which modifies the structure and properties, and results in increased flexibility, improved puncture and impact resistance, improved optical properties and modified sealing performance. By varying the level of vinylacetate (VA) content, the molecular weight, melt flow and additives, it is possible to produce a wide range of copolymers, tailored to meet specific end-use applications. EVA copolymers can be processed easily on most equipment used for packaging applications—blown film, extrusion coated film, cast film and co-extruded films. EVA copolymers are used mostly in multilayer film structures with HDPE, PET, PP, nylons for medical, pharmaceutical, bulk liquid and food packaging. EVA copolymers can also be moulded by injection and compression into liners, wads and seals. They are used in particular as liners in beverage closures for carbonated soft drinks bottles.

2.4.1.2 Ionomers

Ionomers are modified olefin polymers that have particularly useful properties suitable for packaging applications. 'Surlyn' produced by DuPont is the major product used in the packaging industry. It is derived from ethylene–methacrylic acid copolymers and consists of three key elements. The olefin modified with acid groups provides chemical reactivity for adhesion to polar materials, good optical properties and a low melting point. The hydrogen bonds provide high cohesive energy for toughness, melt strength and solvent resistance, and the ionic cross-links enhance these properties. The ionic links are provided by metal ions, usually zinc or sodium. The key properties are excellent sealing at low temperatures, with high seal strength. Peelable seal grades are also available. Other attributes are: toughness, good abrasion, puncture, flex and impact resistance, good chemical, oil and grease resistance, excellent adhesion to other polymers, and adhesion to foil by coating and extrusion coating techniques. Newly developed grades can be very flexible or very stiff.

Ionomers do not provide a good gas barrier, but offer a similar moisture barrier to LDPE. Packaging applications for ionomers include a wide range of multilayer film structures with other polymers, paper and foil for use with various foods, snacks and powders. Ionomers provide excellent sealing and bond performance for flexible composite structures, lidding, seal layers for peelable lidding on glass jars, easy-tear bundling film, towelettes and film for a wide range of food products such as bag-in-box cereals and snacks using form-fill-seal equipment. Also, they are used in bulk liquid packaging with paper, foil and polymer structures for cartons for juice and oils. In combination with high gas barrier polymers they are widely used for food packs where gloss, clarity and good sealing is required (e.g. meat, cheese, fish packs). Ionomers can be injection moulded into thick sections and retain high clarity, with good chemical, scratch and abrasion resistance, as well as stiffness. They are therefore ideally suited for transparent perfume caps for beauty and cosmetics packaging, and some striking examples are available.

2.4.1.3 Polymethyl pentene

'TPX' is the trade name for 4-methylpentene-1-polyolefin (PMP) first produced by ICI and now by Mitsui Petrochemicals. 'TPX' is a crystalline polyolefin like PE and PP but because it has a bulky side group in the chain it does have different properties. It has a high melting point, good temperature resistance, very low density (0.83 g/cm^3), excellent clarity, good stiffness, excellent chemical resistance, but the gas barrier is very poor, and the moisture barrier is lower than that of PE and PP. 'TPX' can be processed by injection, extrusion and extrusion blow-moulding, but has a very sharp melting point and low melt viscosity compared to other polyolefins. It has good alcohol resistance and has been used for cosmetic caps and cases, jars for beauty and personal care packaging, and some limited use in food packaging.

2.4.2 Thermoplastic elastomers

Thermoplastic elastomers (TPE) combine rubber-elastic properties shown by conventional elastomeric materials with the ease of processing of thermoplastics. Unlike classic elastomers, TPEs are not cross-linked covalently, but are reversible cross-linked so can be processed like a thermoplastic. The hard and soft phases can be combined in a number of ways, so considerable overlap between the performance of TPE, elastomers and thermoplastics is possible. TPEs can be classified into two groups on the basis of composition. Polymer blends consist of a thermoplastic matrix (hard phase) in which smaller elastomer particles (soft phase) are embedded. They include TPE-O (olefins non-cross-linked) and TPE-V (cross-linked elastomer). Copolymers consist of elastomeric and thermoplastic sequences in the polymer chain. The main types are TPE-S (styrene block copolymers), TPE-U (polyurethane), TPE-E (polyether ester) and TPE-A (polyether amide). The advantages of TPEs over elastomers are ease of processing and recycling, easy colouring and lower density. Disadvantages are long-term heat stability, long-term loading, compression set and high raw material cost. The least expensive TPE materials are TPE-S and TPE-O types and they account for about 75% of the total market (total: 500 000 tonnes in Europe). TPE-V is forecast to have the highest growth.

TPEs can be processed like standard thermoplastics, often on existing equipment, by injection, extrusion and extrusion blow-moulding. TPE materials have so far penetrated the packaging sector in niche applications only, although there are some food, pharmaceutical and medical applications where TPEs are used in direct food contact. They offer good repeat sterilisability, a low level of extractables, low taint and odour. Applications include: soft touch finish on bottles and cosmetic containers, bags, liners, gas permeable films, and sealing gaskets, wads for caps and closures and cosmetic cases.

2.4.3 Polystyrene copolymers

Polystyrene (PS) as an amorphous homopolymer is brittle and has excellent clarity. Copolymers with monomers like acrylonitrile to produce styrene acrylonitrile (SAN) have much improved impact resistance, temperature resistance, improved stiffness and are less prone to stress cracking in contact with chemicals and food products containing oils, fats, and the fragrances in cosmetics. SAN is also very clear but has a yellowish tint, although glass-clear grades are now available. SAN has similar moisture barrier properties to PS but far superior gas barrier properties. It can be easily processed by most techniques, and can be injection moulded into high gloss caps and closures, compacts, jars used in cosmetics, beauty, and personal care products. SAN can be extruded into tubes, sheet and bottles. The surface finish is normally not as good on extruded components as it is on injection mouldings, especially with transparent grades.

The addition of rubber particles such as butadiene rubber to PS to form high impact PS (HIPS) can also be used to produce a terpolymer of acrylonitrile butadiene styrene (ABS), a high performance polymer with a good balance of thermal, mechanical and impact properties. A small particle size rubber butadiene–acrylate polymer can also be blended into the SAN matrix to produce ABS for improved stress crack resistance applications. ABS like SAN can be easily injection moulded into caps, closures, compacts, lipstick cases for cosmetics and beauty products. Mouldings have good scratch and scuff resistance and high quality surfaces and gloss. High heat grades are possible, as are very high impact grades and high rigidity, so ABS is particularly suitable for electroplating a metallic finish onto the polymer surface for perfume fragrance caps (see Table 2.11).

Table 2.11 Polymers for beauty packaging

	PETG	SAN	PMMA	COC
Transparency (%)	92	90	92	93
Stiffness (Mpa)	2000	4000	3300	2600
Impact strength notched Izod (kJ/m^2)	100	25	20	20
Density (g/cm^3)	1.27	1.08	1.19	1.02

2.4.4 Polyesters

Polyesters which are amorphous (e.g. PETG) contain an aromatic glycol in the polymer chain to prevent crystallisation but retain high clarity. PETG and similar copolymers like PCTG contain different levels of a specific glycol called cyclohexane dimethanol (CHDM). These products are produced by Eastman and have the trade name 'Eastar'. They have high clarity and gloss, low haze, and good mechanical properties, and can be processed by extrusion into films, sheet, tube and profiles and by blow-moulding into bottles, jars and containers. Like PET, these copolymers do not have very good temperature resistance. They have lower gas and moisture barrier properties than PET, but do have good chemical resistance and good scratch resistance, so are suitable for caps, compacts and cases. When moulded into thick sections they retain high clarity and are suitable for cosmetics, beauty and personal care packaging (see Table 2.11).

2.4.5 Polyamides

Polyamides (PA) are also known as nylons and contain an amide group –CONH$^-$ in the chain. Nylons are formed by reacting an amine with an acid, or from lactams (e.g. PA6 from caprolactam). The first polyamide PA66 was commercially manufactured by DuPont in 1938, and called 'Nylon', and the same year I.G. Farbennindustrie first produced fibre-forming PA6 which was called 'Perlon'. PA6 is commonly used in plastics packaging, while PA66 is more widely used in engineering applications, less so in packaging. About 12% of the total tonnage of nylons in Europe (total: 550 000 tonnes) is used in packaging applications.

Nylon 6 and 66 together account for the major portion (45% each) of nylons consumed in all forms of plastics including packaging. There are at least 1500 different nylon polymers and copolymers commercially available, and of these the most widely used in packaging are: PA6, PA6I/6T, PA-MXD6, PA11, PA12. Most nylons are semi-crystalline (50%) and have a crystalline melting point; PA6I/6T is amorphous and has a distinct Tg but no crystalline melting point. Nylons can be processed by most common fabrication technologies into flexible (films) and rigid packaging. Nylon use in packaging is relatively small, the main application being in barrier packaging either alone in monolayer, or together with other polymers in multilayer structures. The main sectors in the consumer goods area are foods and beverages, with some use in personal care and health and beauty packaging.

2.4.5.1 Nylon 6

Nylon 6 has a good range of properties, and is relatively easy to process by the major conversion processes, including injection, extrusion, extrusion blow, and film blowing. The combination of easy processing, excellent surface appearance and easy colouring gives nylon 6 a high level of product and consumer appeal. It is produced from caprolactam monomer by hydrolytic polymerisation in the presence of water. Nylon 6 is a high melting polymer which will readily absorb moisture and when dried ready for processing must be stored in sealed containers to prevent moisture absorption. Nylon 6 homopolymers are available in standard and nucleated grades; the latter are stiffer, have lower impact strength, and are easier to process (shorter cycle times). These are stiff polymers with good impact strength when dry; when conditioned following moisture intake the stiffness will reduce while the impact strength will increase. The chemical resistance of nylon 6 is generally excellent, with good resistance to most food and beverage products, but it is plasticised by water and alcohols. Resistance to fats, oils and grease make nylon 6 ideal for packaging many food products.

The main applications for nylon 6 in packaging are as film in flexible food packs (where it provides toughness, flavour and aroma barrier properties and good oxygen barrier properties), as lightweight blown bags as liners for personal care products in metal aerosol cans (e.g. 'Bican' produced by Crown Cork & Seal), as a barrier layer in rigid HDPE bottles to provide chemical barrier properties, as an outer skin layer on HDPE bottles to provide gloss, in injection closures and components for cosmetics and beauty where high gloss is required.

Nylon is widely used in food packaging, alone and in multilayer structures, usually combined with polyolefins, aluminium foil and PVDC. The main reason for the use of multilayer structures in packaging is to improve barrier properties and reduce cost. The gas permeability of nylon films to oxygen and CO_2 is generally much lower than for polyolefins, but moisture permeability is much higher than for polyolefins. The high temperature resistance of nylons allows them to be used for packaging which requires sterilisation and retorting at 121°C or higher (Table 2.10).

2.4.5.2 Transparent nylons

Many different transparent nylons have been developed, and a few are commercially available and are used in packaging (e.g. toiletries, cigarette lighters, medical and in barrier packaging). The categories of interest for packaging are nylon copolymers, nylon random copolymers, and multi-polyamides. Nylon copolymers with terephthalic acid (T) and isophthalic acid (I) are amorphous, transparent and have good barrier properties, a high Tg when dry, and high stiffness, making them suitable for barrier packaging. Commercial products include PA6I/6T 'Grivory' G21 from EMS Chemie.

2.4.5.3 Nylon blends

There has been some interest in nylon blends for packaging, in particular, blends with polyolefins (PE, PP). Because of their incompatibility, useful blends require a third polymer component to make the nylon compatible with the polyolefin. The blends have improved properties when compared to nylons, such as lower moisture permeability, better chemical resistance to alcohols, toughness and good impact properties. Commercial products include 'Orgalloy' from Elf Atochem.

2.4.6 Other polymers

2.4.6.1 Polymethylmethacrylate

'Acrylics' or polymethylmethacrylate (PMMA) polymers are very high clarity materials more commonly known for their use in lighting, in-store displays and glazing applications. PMMA is however used in packaging and can be processed by injection, injection blow, extrusion and co-extrusion techniques. Special grades with improved impact, heat resistance, high surface hardness, food and medical grades are available. PMMA polymers have good stiffness, good impact, very low scratch and abrasion resistance, and generally good resistance to most chemicals, although some chemicals like alcohols may cause crazing and stress cracking. PMMA, because of its excellent transparency and ease of moulding in thick sections, has been used for injection moulding containers for beauty and cosmetics packaging. Terpolymers of MMA, butadiene and styrene (MBS) are impact resistant, transparent two-phase materials which have good resistance to oils and fats. They can be processed by injection, extrusion and extrusion blow-moulding. Packaging applications include injection moulded jars and blown bottles for cosmetics, beauty packaging, and medical disposables and packaging.

2.4.6.2 Polycarbonate

Polycarbonate (PC) is an amorphous polymer produced by the melt polycondensation of bisphenol A and diphenyl carbonate. PC has excellent transparency, is tough, hard and impact resistant, with a high temperature resistance suitable for steam autoclave sterilisation. PC polymers are easily processed by injection, and by extrusion blow-moulding. Polycarbonate does not have good gas barrier or moisture barrier properties. It has been used for some packaging applications, including extrusion blown and injection blown baby food bottles, returnable milk bottles, beverage bottles and large water containers. Its poor gas barrier properties have prevented any major food or beverage packaging applications. PC plus ABS blends have been used for injection moulded caps which are electroplated for a metallic finish for beauty and cosmetic perfume fragrances.

Although polycarbonate is an extremely tough material, it is prone to environmental stress cracking. Containers have to be annealed after manufacture to remove moulding stresses in order to improve properties.

2.5 New and future developments

2.5.1 Single-site catalyst polymers

Single site-catalysts (SCC) and, in particular, the development of metallocene catalysts has led to the commercial introduction of a wide range of polymers, often with unique properties which are of potential interest for packaging applications.

SCC, of which metallocenes are just one example, differ from multisite catalysts like the Ziegler–Natta catalyst in that they have one active site producing more homogeneous polymers with narrow molecular weight distribution (MWD); copolymers can be produced with very high levels of comonomer which are evenly distributed along the polymer chain. These polymers also have less low molecular weight material, the material which often is the cause of taint and odour problems in packs. SCC polymers potentially will improve this aspect, opening up new markets, such as the packaging of non-carbonated mineral water.

Metallocene PE (mPE) polymers were the first to be developed and represent the widest range of types and grades available from current suppliers. In the HDPE range of mPE polymers there are injection grades suitable for thin-walled containers with very high melt flow and good impact properties: for example, 'Elite' from Dow. In the MDPE range, grades are available for film applications: for example, 'mPact' from Phillips. In the LLDPE range there are a number of grades which have been developed for films with improved properties, including: high stiffness and heat resistance, high puncture resistance, good optical properties, low taint and odour due to low extractables, and low heat sealing temperatures, compared to conventional LLDPE. These polymers are being used in packaging films for foods, stretch film, injection moulded containers: for example, 'Exceed' from Exxon.

Polyolefin plastomers (POP) are a completely new type of PE polymer that have been developed using metallocenes. POPs are PE copolymers with a high comonomer content. They are similar to LLDPE but with the potential of a very low density ($0.86\,g/cm^3$) while retaining high homogeneity. The comonomers used are 1-butene in POP polymers produced by Elenac, called 'Luflexen', and 1-octene in POP polymers produced by Dow (called 'Affinity') and by DEX Plastomers in Europe and Exxon in the US (called 'Exact'). These are soft and very flexible polymers. In film applications POP offers very high puncture resistance, excellent optical properties, optimum ease of sealing, and high gas and moisture permeability (poor barrier properties). POP polymers are suitable for breathable film, laminates in modified atmosphere packaging, and form-fill-seal process packs.

Another new polymer termed polyolefin elastomer (POE) is similar to the POP materials, but has high levels of comonomer, producing soft, very low density, very flexible materials which have potential in packaging applications for liners, seals, wads for caps and closures, flexible bags and pouches. POE polymers are produced by DuPont–Dow Elastomers and called 'Engage'.

Metallocene PP (mPP) polymers are still under development by a number of companies; one commercial range of materials is available from Targor under the trade name 'Metocene'. These are easy-flow polymers which combine stiffness with excellent clarity, low extractables and are especially suited to injection moulding into thin-walled containers, cups, pots and lids for food and beverage packaging. They are being targeted as a possible replacement for PS, SBS and other polymers. Applications include toothbrush and CD cases. mPP polymers are also being marketed for both cast and blown film applications, where their excellent optical properties, ease of sealing and high stiffness make them very suitable for food packaging.

Syndiotactic PP (sPP) is being developed by Fina. It has a much lower melting point than isotactic PP (iPP), is more flexible and has good clarity and low haze. sPP appears to be difficult to process and control, and this is delaying its commercialisation.

Cyclic olefin copolymers (COC) have been researched for some years, and with the advances in catalyst technology in particular of metallocenes, COC polymers are now viable for large scale production. COC polymers are copolymerised from ethylene and 2-norbornene, both readily available monomers, using metallocene technology by Ticona to produce 'Topas' grades of materials. These are amorphous, highly transparent copolymers with low density, high stiffness, excellent moisture barrier properties, and high temperature resistance. COC polymers can be sterilised using a steam autoclave and by gamma radiation so that these properties, together with excellent chemical resistance, make them very suitable for medical and pharmaceutical packaging applications, and offer potential for food packaging. COC polymers like some other amorphous polymers are susceptible to certain organic chemicals, oils and fats, and must be tested before final selection for a particular packaging application. The other producer of COC polymers is Nippon Zeon in Japan with a non-metallocene catalysed polymer called 'Zeonex'. This is a cyclic olefin polymer produced from ethylene and dicyclopentadine monomers. The polymer has similar properties to 'Topas' and is being targeted at similar markets, which include packaging applications. Nippon Zeon have recently introduced a lower cost version called 'Zeonor' which is also of potential interest as a packaging material.

Dow has recently announced their 'Index' range of ethylene styrene interpolymers (ESI). This development has been possible using Dow's 'Insite' single-site catalyst technology, which enables predictable and consistent copolymerisation, allowing ethylene or styrene composition to be varied from less than 25% to over 80%. EIS polymers offer the toughness and solvent resistance of PE along with the stiffness, ease of forming, and appearance of PS. 'Index' polymers will be available in two series, distinguished by the phase structure, either semi-crystalline elastomer (E series) high ethylene content polymers, or amorphous semi-rigid (S series) high styrene content polymers. Potential packaging applications include: soft touch finish, beauty containers and personal care bottles.

Looking ahead to further advances that can be expected, in current packaging markets single-site catalyst polymers have yet to make a major impact in areas apart from films where they are currently being used. Some polymers are being evaluated for injection moulding of thin-walled containers, but so far not into caps and closures. Blow-moulding grades are not yet available because it is more difficult to process narrow MWD SSC polymers, but these should be developed in the future. Flexible mPE and mPP grades will continue to be developed, and find uses in packaging applications. Will a high gas barrier SCC polymer be developed to rival PET and other barrier polymers?

2.5.2 Barrier developments

New developments in barrier polymers and technologies currently seem to be announced almost weekly. The key issues are: 'How best to improve the gas barrier of PET' and 'Will PET be suitable for packaging beer?'. The options available are numerous and varied. Multilayer co-injection technology is being developed commercially by a number of companies. Multilayer preforms can be produced in multicavity tools using EVOH or N-MXD6 as the barrier layer, and oxygen scavengers and absorbers can also be incorporated into the layer. Various coating processes are being developed. PPG offers an epoxy-amine coating material called 'Bairocade' which can be externally coated onto bottles to give a very high gas barrier clear coating suitable for packing beer. 'Bairocade' technology has been commercialised by Containers Packaging for monolayer PET beer bottles in Australia, and Graham Packaging in the US for juices in PET. Crown Cork & Seal is developing its coating technology under the trade name of 'Starshield'.

A new exterior coating process has been developed jointly by the Coca Cola Company working with Krones and the University of Essen called 'BEST PET'. A clear silica, silicon oxide (SiO_x), is applied to the external surface of PET bottles using a patented high vacuum plasma process, based on the use of physical vapour deposition (PVD). The technology is claimed at least to double the gas barrier properties, which could potentially reduce the thickness of PET bottles. The coating is claimed to be clear and colourless. A second process being developed by TetraPak, called 'Glaskin', plasma coats the inside of PET bottles with SiO_x to produce a clear coating 0.2 μm thick. It is also stated to improve the gas barrier properties by a factor of at least two.

A further plasma deposition process has been developed by Sidel called 'ACTIS'. This time amorphous carbon is deposited onto the internal surface of PET bottles with a thickness of 0.1 microns, using acetylene gas and microwave energy directed at the bottle. PET bottles can be produced with a 30 times better (oxygen) and 7 times (carbon dioxide) gas barrier. The coating is clear but has a yellow tint, making it suitable for beer, which is usually packaged in brown bottles. This coating technology offers one of the best systems that could rival the gas barrier performance and economic costs of glass and metal cans packaging. A similar barrier coating technology developed by Kirin and Mitsubishi Shoji Plastic uses plasma vapour deposition of an internal coating of a thin layer of diamond-like carbon (DLC). This is claimed to give excellent gas barrier improvements to PET bottles.

Another development for barrier PET bottles is external over-moulding of a barrier layer onto the injection preform, prior to stretch blow-moulding into a two-layer PET bottle. TetraPak has developed this technology together with another company. The barrier layer could be selected from a range of polymers and the layer thickness adjusted depending on the barrier performance required.

Oxygen scavenging or absorbing materials remove oxygen in the headspace of the pack, from the product, or in the walls of the pack in contact with the product. These have been developed by a number of companies over the years. BP Amoco has developed 'Amosorb', a polyester copolymer containing an iron salt which acts as an active barrier layer in PET bottles, is compatible with PET and does not delaminate. The Metal Box Company, some years ago, developed its 'Oxbar' scavenging technology, using a blend of PET with nylon N-MXD6 containing a cobalt salt, as the barrier layer in PET bottles. This technology is now being further developed by Crown Cork & Seal. Other companies active in this technology are Toyo Seikan with 'Oxyguard', an iron salt scavenger, and Chevron is working on polyolefin scavengers.

Inorganic materials like silicates based on montmorillonites are nanoclays which are used to produce nanocomposite materials. These nanoclays can be blended at low levels (5%) with polymers to produce high gas barrier nanocomposites. The silicates produce a lamellar (plate-like) structure which extends the diffusion path in packaging materials, so improving the gas barrier properties. The nanocomposite can be in a separate layer or a blend in the monolayer package structure. Nanocor is one of the leading manufactures of silicates, and is developing nanocomposite materials together with a number of companies. Most of the development so far has focused on nylon based nanocomposites, and these have been incorporated as a separate layer with nylon 6 films to improve gas barrier. Nanoclays are very small particles which do not interfere with light transmission through the pack, so clarity can be maintained. The level of haze is acceptable for amber bottles, but not presently for clear bottles.

In films puncture resistance and stiffness is improved and the materials can be processed on existing equipment, with the potential for thinner barrier layers in both films and rigid packs. Nanocomposites offer another alternative for gas barrier improvement in PET and possibly other polymers. Eastman is working with Nanocor on nylon based nanocomposites for use in multilayer PET bottles. Eastman will commercialise its 'Imperm' nanocomposite materials during 2000.

The Dow barrier thermoplastic epoxy polymer 'BLOX' is an amorphous material which can be used as a barrier layer in multilayer PET structures, and as a coating for bottles. Tailored comonomer design offers a range of grades from medium to high barrier. Dow believes 'BLOX' will be more cost-competitive than alternative barrier polymers in barrier structures for beverage packaging of beer and juices. The polymers have excellent transparency with low haze, excellent gloss and good colour, and excellent adhesion to PET.

The excellent gas and moisture barrier properties of liquid crystal polymers (LCP) which are highly aromatic polyesters, has so far not been exploited for packaging, mainly because of economics; LCPs are very expensive polymers, and technically it is difficult to process preforms containing LCP by stretch blow-moulding. Blends with PET have been developed by Superex and bottles produced from these have good gas barrier properties, but are opaque because of the multiphase structure. A clear LCP would be a very interesting prospect for packaging.

Increasing the barrier properties of PET bottles by any technique will potentially open up the market for the packaging of many oxygen-sensitive products like beers. To extend this penetration further, bottles will need to withstand pasteurisation, and to achieve this, PET will need to be heat set to increase the crystallinity and temperature resistance, and also to improve its pressure resistance. Plastic closures are being considered for pasteurised beers, and these will also require to have high temperature resistance and the ability to withstand high internal pressures. In addition plastic closures and plastic liners used for pasteurised beverages will need to have improved gas barrier performance to match that of the bottle. Many premium brands of beer are tunnel pasteurised at temperatures between 60 and 70°C for 5 min or longer to guarantee shelf-life and quality of product.

Future development of polyesters will depend on the economic costs of manufacture of high barrier polymers, in particular, naphthalate-containing polymers. Shell has recently introduced a polyester called polytrimethylene terephthalate (PTT) for packaging films with a slightly better gas barrier than that of PET. The next stage of their development is to produce the chemically similar polytrimethylene naphthalate (PTN) which has much better gas and moisture barrier properties than either PET or PEN.

2.5.3 Biodegradable polymers

Biodegradation is the breakdown of organic material by the action of microorganisms, usually by composting in the soil. Biodegradable polymers are being developed by a number of companies, essentially using two different types of base materials—renewable raw materials and base petrochemicals derived from oil and gas. These polymers are yet to become significant in packaging applications, but the prediction is that in 5–10 years this could change, and volumes increase.

Materials available today can be produced from a wide range of technologies from renewable resources (animal or vegetable origin), or non-renewable fossil resources. Biodegradable polymers have been available and used in packaging for many years; for example, cellulose and its copolymers in film form have been used for confectionery and dry non-fat foods. More recently Danone has used poly(lactic acid) (PLA) to thermoform yoghurt tubs, replacing PS. Other applications include milk containers, flexible coffee packs, confectionery packs and cereal wraps.

PLA, the most interesting polymer, is being developed by Cargill Dow Polymers with NatureWorks, and by Mitsui Chemical with Lacea. The properties of PLA, which is derived from corn, starch and other plant materials, are similar to PS and PE. It has high clarity and gloss, high stiffness and strength, and good flavour and aroma barrier properties.

Biodegradable polymers enable a materials cycle to be set up which is similar to that existing in nature. It creates a potentially cheap access to plant-based raw materials and a cheap method of disposal by composting.

3 Structural design, CAD/CAM, performance and finite element analysis

H.A. Hughes

3.1 Introduction

Modern design and manufacturing depend heavily on the use of computers and software. Computer assisted design (CAD) harnesses the power of computers to speed up and improve the process of creating a design. In particular, the ability to modify an existing design drawing quickly and easily makes it possible to examine more alternatives in an effort to find an optimum design. Computer assisted manufacturing (CAM) uses the information developed by the CAD system to make prototypes, plan production, estimate time requirements and monitor manufacturing systems. The CAD file can be processed into codes which directly control computer numerical controlled (CNC) machines that manufacture products and components, or it can be used indirectly to make molds or forms to be used in subsequent manufacturing operations. Finite element analysis (FEA) can be used to evaluate any system that can be described by a set of differential equations to determine stress distributions, patterns of heat flow, movement of liquids or similar measures of performance. The information from the FEA can be used to guide subsequent design iterations, in the process of searching out the optimization of a design.

3.2 Design

Any product*, whether a small consumer item, such as a plastic dish or bottle, or a large industrial package, such as a returnable plastic tote, starts as a design. Design is the process of planning all details of a product. The details include material selection, dimensions, shape, manufacturing method, distribution method and any other relevant information. The design process is a disciplined effort that maintains the opportunity for the designer(s) to be creative in the solution of the various problems which must be addressed.

The process of design does not produce a product in the usual meaning of the term. Rather, the design process produces a plan or template that can be used to replicate a particular product as many times as is required. In the case of

*For the purpose of this chapter, the term product can be taken to be either a 'conventional' plastic consumer product or a plastic package.

plastic packaging, the replication process is usually done many times. A popular package, such as a 16 oz soft drink bottle, a 2 l PET bottle or a plastic closure, may be replicated millions (occasionally billions) of times.

Design details can be particularly important when many units are to be manufactured. For example, the saving of a small amount of material per container, when replicated millions of times, can amount to a huge overall cost saving for the manufacturer.

Design has a direct impact on the performance of the package. In particular, the performance of packages used to contain and protect food, pharmaceutical products and hazardous materials is critically important. Inadequate food packages can allow leakage, quality deterioration or loss of wholesomeness of food products. Inadequate pharmaceutical packages may cause the product to lose efficacy. Inadequate packages for hazardous materials may fail, exposing workers and innocent bystanders to unanticipated dangers.

Package design includes the decoration, labeling, and other aspects of the appearance of a package. These are important considerations in the overall design of packages, but are not covered in this chapter. Rather, the discussion focuses on structural considerations, such as the package shape, dimensions and strength.

3.3 CAD/CAM

The acronym CAD can be given two meanings: computer assisted drawing or computer assisted design. The two terms are often used interchangeably, but there is an important difference:

- *Computer assisted drawing* is a system that uses the capabilities of a computer and an appropriate software package to aid a designer to prepare a plan for a product. The software can usually display the object in various ways to help the designer to understand its shape and size and possible fits with other objects, but there is no inherent ability to analyze the potential of the package to withstand the impacts and loads imposed on it during the filling, handling and distribution operations.
- *Computer assisted design* is a more sophisticated approach, using software that can analyze the behavior of an object under dynamic or static loads. It can analyze the results of changes in design dimensions and can often display the motion, deflection, vibration, and other behavior of an object as it responds to simulated loads. The computer assists the designer to create, analyze and optimize the design.

The acronym CAM stands for computer assisted manufacturing: the use of computers and software to control and manage various manufacturing activities. Generally, the computer file produced during the design process can be transmitted to a computer that controls a milling machine or other equipment used to make molds or forms. Rapid prototyping (discussed in another chapter) is a specialized application of CAM.

CAD/CAM integrates the design and manufacturing processes. Traditionally, manufacturing was essentially disconnected from the design process in many companies. Modern systems, however, which use the design, communication and control capabilities of computers, can design a product, produce prototypes, and put the product into production more quickly, with a lower incidence of errors and at lower overall cost.

The following section discusses CAD as computer assisted drawing. The discussion of CAM and the use of computers for design analysis follows in later sections.

3.3.1 Computer assisted drawing

A minimal CAD system includes a computer, a monitor, a pointing device, an output device, such as a printer or a plotter, and an operator. This minimal setup is suitable and efficient only for individual designers working in small businesses. More commonly, there is a more elaborate system, consisting of a networked arrangement of computer workstations connected to a central server. Each workstation consists of a local computer, a large high resolution monitor, and a pointing device. The server handles storage and management of design files and the databases of component designs, specifications and other information needed by the designers. The server also controls the output devices, usually consisting of a set of different types of printer and plotter. There may be other specialized equipment, such as devices that measure objects and automatically feed the data into the computer for subsequent manipulation.

3.3.1.1 Types of CAD systems

CAD systems can work in either two or three dimensions. All CAD systems include a basic set of tools for constructing a drawing from a set of fundamental components. Typical components include: straight lines, multilines, arcs, circles, ellipses, text, polylines, polygons and dimensions. Most commands can be accessed by using a pointing device to select an icon, either on the monitor screen or on a tablet. Commands often include an array of specialized versions of the same instruction. For example, one popular CAD program provides more than ten ways to draw an arc.

CAD systems also include an extensive set of editing tools. Editing tools allow components of the drawing to be modified in various ways. A typical set of editing tools includes: copy, move, mirror, rotate, extend and trim. The editing tools are often more useful to a skilled operator than the basic tools, because they allow time to be saved. The command 'Array' is an example. The

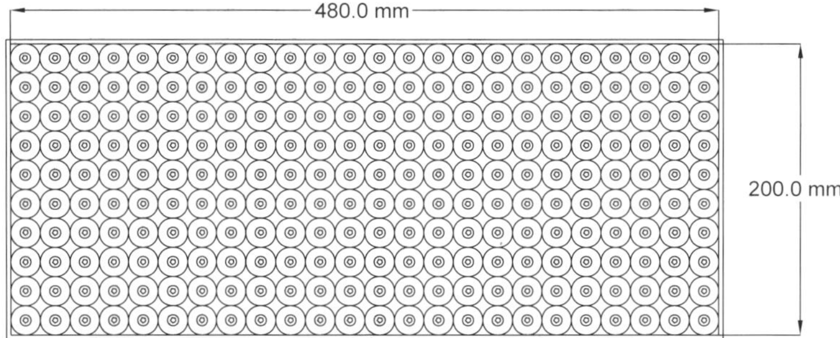

Figure 3.1 A set of 240 similar objects can be drawn in the desired arrangement of rows and columns by executing a single command after the drawing of the initial object is constructed. The figure shows an arrangement of 240 plastic bottles in a carrying tray (10 rows, 24 columns). All bottles are 20 mm in diameter. The plastic corrugated tray is 3 mm thick.

set of 240 similar objects shown in Figure 3.1 were drawn by executing the 'Array' command one time after the initial object was drawn.

3.3.1.2 Two-dimensional CAD

Two-dimensional CAD systems are best suited to making drawings of flat objects, such as box blanks. A box is, of course, a three-dimensional object, but it is manufactured from a flat piece of paper, plastic corrugated board or paperboard. A box plan, as shown in Figure 3.2, can be easily and appropriately drawn on a two-dimensional CAD system.

A two-dimensional CAD system can be used to produce three-dimensional drawings of objects, using the projection approaches that were historically used

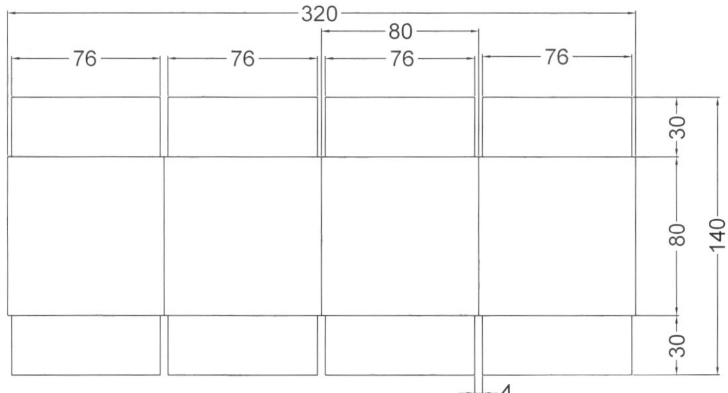

Figure 3.2 Two-dimensional drawing of a plastic corrugated box blank. All dimensions in mm.

by draftsmen working with pencil and paper. However, such drawings lack many of the features of three-dimensional CAD drawings.

3.3.1.3 Three-dimensional CAD

Three-dimensional CAD is a powerful extension of two-dimensional CAD, including a set of additional commands that facilitate the construction and manipulation of three-dimensional objects. Of course, the objects are presented on a two-dimensional screen or paper. A three-dimensional system can move and rotate objects on the monitor screen, allowing the designer to examine the object from various viewpoints. One such view would be the same as the object would appear if drawn on a two-dimensional system.

The process of constructing the drawing of an object on a three-dimensional CAD system can be confusing to an inexperienced designer. To assist, the CAD screen can display the appearance of the object being drawn in many views. A typical set of views includes: front, top, bottom, back, left side, right side and orthographic. Lines and other drawing components can be added or modified in any view, with the results shown in all views. This is helpful when identifying and correcting errors.

The set of drawing tools or components used to construct a design by a three-dimensional CAD system is similar to the set used by two-dimensional systems. However, the editing commands are more extensive and more powerful. One such command, frequently called ‘Extrude’, allows a circle to be turned into a cylinder. Another useful command is ‘Blend’, which allows two surfaces to merge smoothly into a single surface. For example, the steps used to construct the basic shape of a long-necked bottle can be drawn as follows:

1. Draw a circle of the diameter of the body of the bottle.
2. Extrude the circle to form the cylinder which is the body of the bottle.
3. Draw a circle of the diameter of the neck of the bottle.
4. Extrude the second circle to form the neck of the bottle.
5. Move the drawing of the neck into position relative to the body.
6. Blend the two cylinders, forming the complete bottle, as shown in Figure 3.3.
7. Add the bottom of the bottle, the finish and other details.

It should be noted that step 5 allows the neck of the bottle to be placed in several positions relative to the body. For example, as shown in Figure 3.3, the neck can be at the center or it can be offset any distance up to the radius of the body.

Three-dimensional shapes can be formed in other ways. For example, if a line defining the outside of a container is drawn, it can be ‘rotated’ about a center line to form a cylindrical solid.

Three-dimensional CAD programs typically include these and other commands that allow shapes to be formed and manipulated. The designer can observe the three-dimensional object from any direction, as shown by the samples in

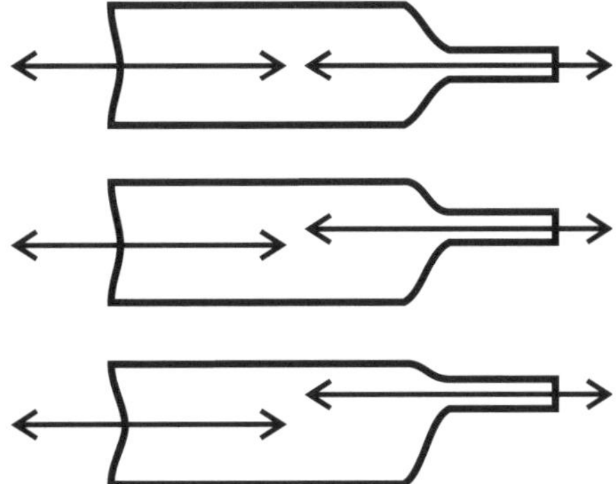

Figure 3.3 Examples of positions for the neck of a bottle. Petroleum products are commonly packaged in bottles with off-center necks.

Figure 3.4. The objects can be shaded to assist the designer to visualize the appearance of the object. A realistic understanding of the geometry can often be a problem when the object is presented as a wireframe, as shown in Figure 3.5.

The three-dimensional drawing file can be used to produce various kinds of output. The most obvious is a printed version. Printed copies can be used as aids in discussions with clients, other design personnel, management and others concerned with the design. By using e-mail, the design can be transmitted long distances, even to other countries, before being printed. Printed copies can also be filed as a permanent record.

Other applications do not require a printed copy of the design. In particular, the drawing file can be used with CAM processes to manufacture the product, as discussed in the next section.

3.3.2 Computer assisted manufacturing

Computer assisted manufacturing is the use of a computer and appropriate software to generate manufacturing oriented data that can be used to control manufacturing machines and processes. The control functions extend beyond the simple machine operations that are involved in the manufacturing of parts or components. The following set of functions are involved in CAM operations:

- the computer can prepare control programs for automatic machine tools
- the computer can prepare a listing of the steps in the operation sequence required to produce a particular product or component
- the computer can determine the time needed for a particular production operation
- the computer can determine an appropriate schedule for meeting production requirements

- the computer can be used to determine when to order raw materials and purchase components and how many should be ordered to achieve the production schedule and
- data collected from the factory activities can be compared to design data to determine the status of the manufacturing activity

Figure 3.4 The ability to rotate a drawing to show an object from different directions is a powerful feature of three-dimensional CAD programs.

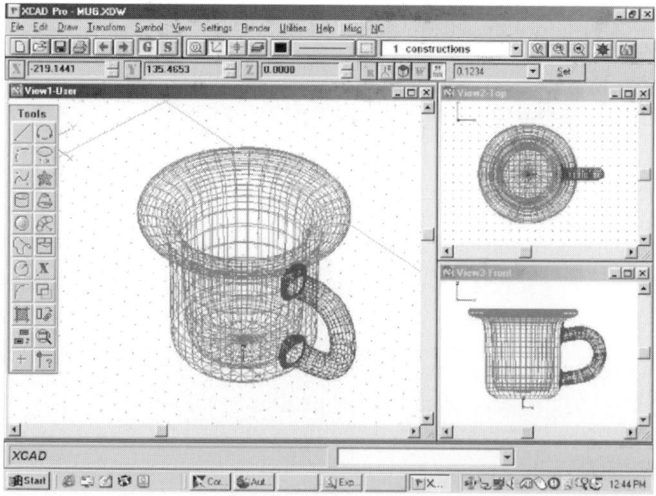

Figure 3.5 A three-dimensional object shown in wireframe, a common modeling technique. The screen is set to present three separate views of the object.

The simplest and easiest to understand CAM operation is the direct control of a machining operation. Software interprets the design and prepares a set of coded instructions to control a CNC machine tool, such as a milling machine. The instructions, based on an X–Y–Z coordinate system, reduce the machining operation into a sequential set of instructions that move the material from place to place under the cutting tool, tracing out the pathways needed to make an object of some type from a block of material. During the preparation of the control codes, additional information about the material and the tools must be provided. The material which is to be machined is specified. The material to be shaped may be a metal, such as aluminium or steel, an epoxy or wax-based material for a prototype, or hardwood for a thermoform mold, as shown in Figure 3.6.

Cutting tools are usually selected from a database of available tools. The milling cutter tool diameter and length, number of cutting flutes, and end shape must all be selected. The tool (rotational) speed must be set. Proper tool speed depends on material properties, depth of cut, and speed of machine carriage movement. The selection of tool and speed changes as the cutting process moves from the initial rough cuts to the finish cut. In the final smoothing operations, the tool usually indexes a smaller distance and the tool rotates at a higher speed. The speed of the feed rate (carriage speed) may also be set to a lower value.

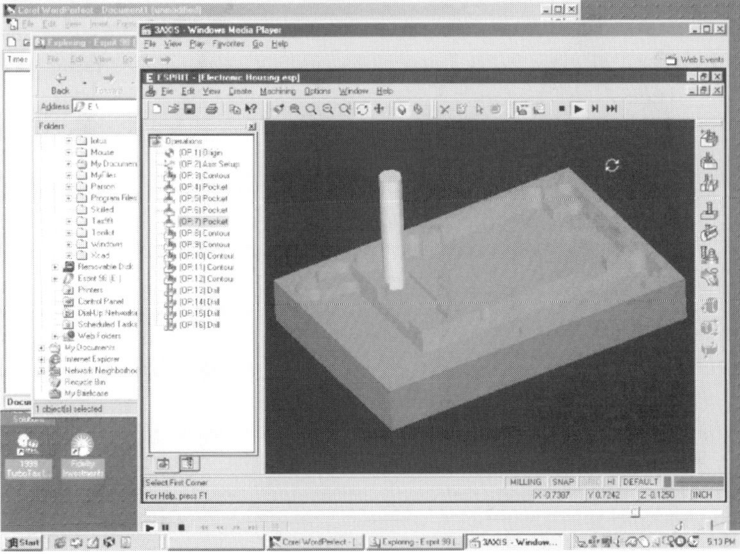

Figure 3.6 Simulation of a milling operation. The simulation is shown as an animation on the screen of the workstation.

Table 3.1 lists a sample of codes and machine movements supported by one CNC milling machine. The CAM software produces a list of codes which are executed in order by the machine tool. The cutting operation always begins with the tool at a home position relative to the material on the carriage. The initial commands will set the rotational speed of the tool and move the carriage so that the tool comes into contact with the material. Then, as the remaining CAM instructions are executed, the tool moves up and down in the Z direction, while the carriage moves back and forth in the X and Y directions. Care must be taken to ensure that the tool does not come into contact with the carriage or any other part of the machine and that it is appropriately in contact with the material. As a means of checking that these requirements are satisfied, there are software programs that can simulate the cutting operation. The simulation appears on the screen, as shown in Figures 3.6 and 3.7, and can be observed by the operator. The speed of the simulation can be set to run faster than the operation would proceed in real time, saving operator time. During this process, the computer will also calculate the time needed to perform the machining operations in real time, providing data needed to predict production rates.

Table 3.1 A sample of the set of G codes and M codes supported by one CNC milling machine

G code	Description
G0X..Y..Z..	Rapid movement
G1X..Y..Z..	Linear movement at feed
G2X..Y..Z..	Clockwise circular interpolation
G1X..Y..Z..	Counter clockwise circular interpolation
G38X..Y..Z..O..D..	Rectangular pocket
	'X..Y..Z..' is the opposite corner of the pocket
	'O' is the amount to offset each successive rectangle
G50X..Y..Z..	Position reset
G70	Set units to inches
G71	Set units to mm
M0	Programed halt
M2	End of program
M3	Start spindle forward
M5	Spindle stop
M6	Tool change

Some objects, such as the item in Figure 3.8, can be made by a machine with a single tool spindle that operates in the vertical position. Other objects, such as the item in Figure 3.7, must be machined on several surfaces. This requires that tools be turned to a horizontal position or that the block of material be repositioned. Moving the material can be quite difficult, depending on the design, so it is preferable to have a machine with multiple spindles or with spindles that can be positioned to work in other directions in addition to vertical.

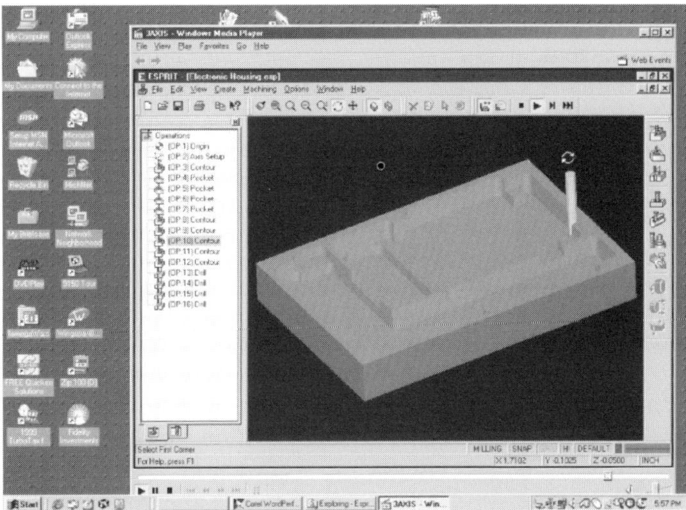

Figure 3.7 A later view of the simulation showing the form of the completed cavity. The tool has been changed to a smaller diameter unit used to make the mounting holes.

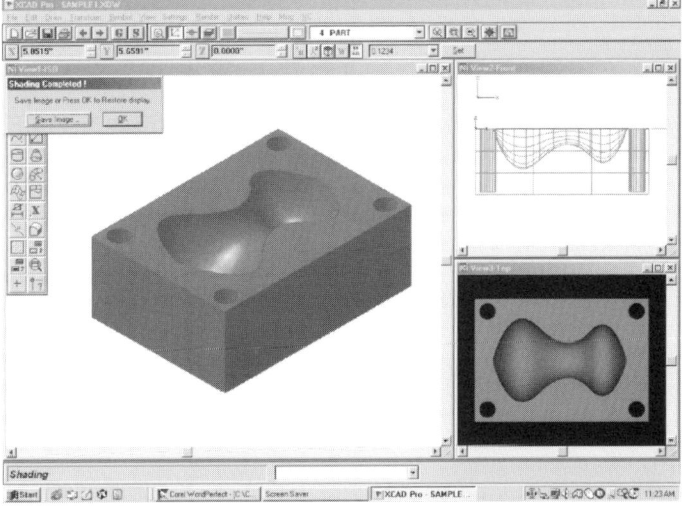

Figure 3.8 A 'peanut' shaped mold that can be manufactured by a single spindle milling machine. There would probably be a tool change during the process.

As mentioned previously, bottles and other forms of plastic packaging are manufactured in large numbers. The CNC machines are therefore generally used to make molds and forms which are subsequently used repeatedly to make individual packages. CAM techniques are used to manage the production process, for example, in setting production schedules, monitoring processes, and reporting to management. Historically, the process of design included extensive testing programs to evaluate the design and determine whether changes should be made. The testing was both expensive and time-consuming, and often did not yield the specific detailed information needed to improve a design. This situation has changed since the development of the technique of finite element analysis (FEA).

3.3.3 Finite element analysis

Finite element analysis is a powerful numerical procedure for solving engineering problems. The method has been in use since the mid-1950s. The first uses were for analysis of structural elements, usually in aircraft, but the versatility of the method and its underlying mathematical basis were soon applied to non-structural problems. Today, engineering applications range from analysis of automobile frames to heat transfer processes involved in the cooling of nuclear power plants and the flow of water or pollutants through soil.

Packaging applications of FEA include the study of stresses in the side walls of paperboard packages/carriers for metal cans of soft drinks and evaluation of the stresses in blow-molded plastic bottles. Stresses in bottles are produced by a combination of internal pressure, compression resulting from stacking and impacts and vibrations which occur when the product is handled.

A complete appreciation of the finite element analysis method requires a strong understanding of advanced mathematics, particularly determinants and matrices. The topic of FEA is too large and comprehensive to be covered adequately in this short chapter, so this presentation covers only the basics of the FEA approach, including the relationship with CAD. Fortunately, there are many excellent reference books on the topic and software to carry out the analysis is available. This software works well and can simplify the work involved in setting up and solving the complicated sets of equations which are involved. However, a cautious approach is in order. Anyone who is not familiar with the FEA approach should arrange for a good training course and should develop the expertise needed to solve a set of simple and advanced problems before attempting to apply commercial software to a problem.

The finite element method approximates a continuous function (such as temperature, pressure or displacement) by a discrete model that is composed of a set of piece-wise continuous functions which are defined over a finite number of subdomains called elements. As shown in Figure 3.9, the elements can be one-dimensional, two-dimensional or three-dimensional in shape, depending on the particular problem. One-dimensional elements can be straight or curved lines. Two-dimensional elements can be flat or curved surfaces, generally triangular or quadrilateral in shape. Under some circumstances, the edges of the triangle or quadrilateral can be curved. Three-dimensional elements are variations of

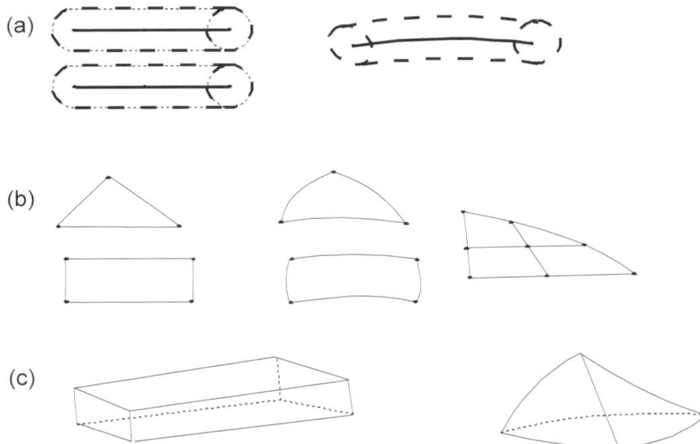

Figure 3.9 Element samples of (a) one-, (b) two- and (c) three-dimensional types.

the two-dimensional elements, namely the tetrahedron and the parallelepiped. Again, under some circumstances, the elements can have curved sides. It can be seen that the system is very flexible and can be applied to a broad range of problems. In fact, present day applications of FEA include all physical problems that are governed by differential equations. Some of the advantages of FEA are:

- the material properties in adjacent elements do not have to be the same. This allows the method to be applied to bodies composed of several materials;
- irregularly shaped boundaries can be approximated using elements with straight sides or matched exactly using elements with curved boundaries. The method is not limited to 'regular' shapes with easily defined boundaries;
- the size of elements can be varied. This property allows the element grid to be expanded or refined as the need exists;
- boundary conditions such as discontinuous surface loadings present no difficulties for the method. Mixed boundary conditions can be handled easily;
- a single general computer program can handle all problems in a particular subject matter area. For example, a general computer program for axisymmetric heat transfer can solve any problem of that type. The ready availability of powerful computing equipment makes the application of FEA to any suitable problem possible.

The relationship between CAD and FEA is clear. A CAD program is used to develop a file that includes all of the geometric information needed to describe or specify an object. This file can be used by an FEA program to lay out the elements and guide the operator to develop the set of equations which can then be solved by the computer.

The FEA technology has two main applications when working with plastic packaging. During the design process, FEA can be used to analyze and predict performance of an object of a particular design, such as a bottle. The results of the analysis can then be used to guide another iteration to change (improve) the design, such as removing material to save costs. This process can be continued until the design has been optimized.

A second use of the method is forensic. If a package fails in use, the design can be analyzed to determine the cause of the failure and then redesigned to strengthen or otherwise improve performance.

Modern design processes make use of all of the types of technology that have been discussed in this chapter. The result has been an improvement in quality of design, performance of plastic packages and economy.

Further reading

1. Segerlind, L.J. (1976) *Applied Finite Element Analysis*, John Wiley and Sons, New York.
2. Stark, J. (1986) *What every engineer should know about practical CAD/CAM applications*, Marcel Dekker, New York.
3. Minitech Machinery (1996) A guide to the operations of the Minitech CAD/CAM/CNC program ADCA-01 Version 1.00, Atlanta.
4. Irons, B. and Shrive, N. (1983) *Finite element primer*, Halstead Press, New York.
5. Akin, J.E. (1986) *Finite element analysis for undergraduates*, Academic Press, Harcourt Brace Johanovich, London.
6. Champion, E.R. and Ensminger, J.M. (1988) *Finite Element Analysis with personal computers*. Marcel Dekker, New York.
7. Autodesk (1997) AutoCAD R14. Sausolito, CA.
8. Autodesk (1998) AutoCAD LT97. Sausolito, CA.
9. DP Technology (1999) Esprit98. DesPlains, IA.

4 Rapid prototyping and tooling

M. Naylon

4.1 Overview

This chapter covers rapid prototyping methods such as concept modelling, stereolithography, selective laser sintering, laminated object manufacturing and fused deposition modelling. The chapter also discusses vacuum casting in silicone rubber tooling, and Keltool for production tooling.

4.2 Rapid prototyping

The manufacturing of an object invariably starts with a design. This design can be a simple freehand drawing, an isometric drawing, a set of engineering drawings, or—increasingly in today's manufacturing environment—an image captured in a 3D CAD file.

From these initial expressions of the design will come perhaps a model made by a pattern maker or a hand-made prototype fabricated from sheet plastic glued and welded together or a block model carved from rigid foam. Sometimes the manufacturer will proceed straight to tool making, taking the risk that the design is acceptable, both aesthetically and functionally. Many will argue that they cannot spare the time or money to spend in the early stage of the product's development, but in fact, money and time spent at this stage is well spent. It costs far less to change a feature on a prototype than to make changes once the tool has been made.

The first stage in rapid prototyping can be called virtual prototyping. In this instance, the design exists only as an image in a 3D CAD file. It is now that changes to the design, styling, dimensions and features are the least expensive to carry out. If the design calls for a lid to fit a bottle, for example, many of the commercially available CAD programs allow this to be done 'on screen'. It may be that the product being manufactured has to fit a standard pallet. This can also be tried on screen to get the correct external dimensions for the product. Using one of the several rapid prototyping processes now available, it is possible to turn that 3D CAD image from a virtual prototype into a real one.

The rapid prototyping process starts with a computer generated 3D model. The CAD software provides a medium for defining geometry and verifying aesthetic and dimensional characteristics at the modelling stage. The CAD file of the model is electronically transmitted, usually in the form of an STL file, to a rapid prototyping system. Several types of system are available for rapid prototyping, and each has its own physical modelling technique.

System choice depends on factors such as model accuracy, cost, model material, type of model, and to what use the model will be put. Rapid prototyping today is taken to mean the manufacture of objects directly from 3D computer generated images by additive layer building methods. This is distinctly different from the traditional manufacturing methods which can be thought of as subtractive (machining, turning and other methods where material is removed) or formative (pressing, stamping or otherwise mechanically deforming planar materials).

4.3 Concept modelling

In many cases in the early stages of product development, it is useful to be able to examine a variety of concepts before the final design, with all its detailed dimensions and features, is fixed. Concept modelling is a recently available method to enable the rapid and inexpensive evaluation of several variations of a concept to be made before proceeding to a functional prototype.

Multi-jet modelling builds three-dimensional objects layer by layer from a wax-like thermoplastic. The objects can be solid or hollow but are not regarded as functional prototypes. They are however considerably less expensive than more traditional rapid prototypes or even NC machined and handmade block models.

Other concept modelling systems produce objects in a thermoplastic polyester by the fluid deposition modelling process, described later, or in a starch-based material by spraying a binder onto the surface of a bed of powder, thereby 'fusing' each layer.

In all cases, be it wax, thermoplastic polyester or the modified starch, objects made by these processes are of limited use, somewhat inaccurate and of rather poor surface finish. However, because of their relatively low cost several iterations of a design can be examined for very little cost.

4.4 Stereolithography (SL)

The stereolithography process was developed by 3D Systems, California. The SL method builds models when a laser cures the prototyping material, in layers 0.25 to 0.025 mm thick. The material used is a photosensitive epoxy resin.

The process starts with CAD data. A stereolithography apparatus (SL) is the machine that receives design data from the CAD file and electronically slices the data into thin cross-sections. At this point, an ultraviolet laser traces each successive cross-section of the object on to the surface of a vat of photosensitive resin. The liquid plastic hardens only where touched by the laser beam.

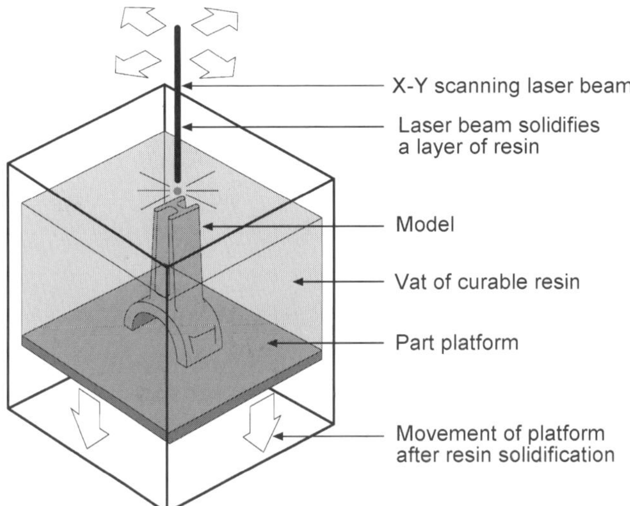

Figure 4.1 Stereolithography.

The object, which is attached to an elevator platform by a support structure, is lowered into the vat and a coater blade sweeps across the hardened surface, thus ensuring the exact thickness of resin is in place for the next layer (Figure 4.1). The next layer is traced on the newly spread liquid, whereupon it hardens as before and bonds to the layer below. The process repeats for each layer until the object is formed. On completion the elevator moves upward to the top of the vat, allowing excess resin to drain from the object.

The object is removed from the elevator platform and separated from the supports before being washed, dried and given a final post cure in a UV light chamber. Finishing consists of carefully working the surface with scrapers, files and various grades of wet and dry abrasive paper in order to remove build lines, support vestiges and give the object a uniform surface finish. Finally the object can be glass bead blasted or polished to a high gloss finish.

4.5 Selective laser sintering (SLS)

The selective laser sintering process (Figure 4.2), offered by the DTM Corporation, Texas, builds prototypes in a wide range of plastics and other materials. It does so by fusing together particles of heat fusible material with a carbon dioxide laser. The bed of powder is held in a chamber at a point below the powder's melting point. When the focused laser beam draws the object's

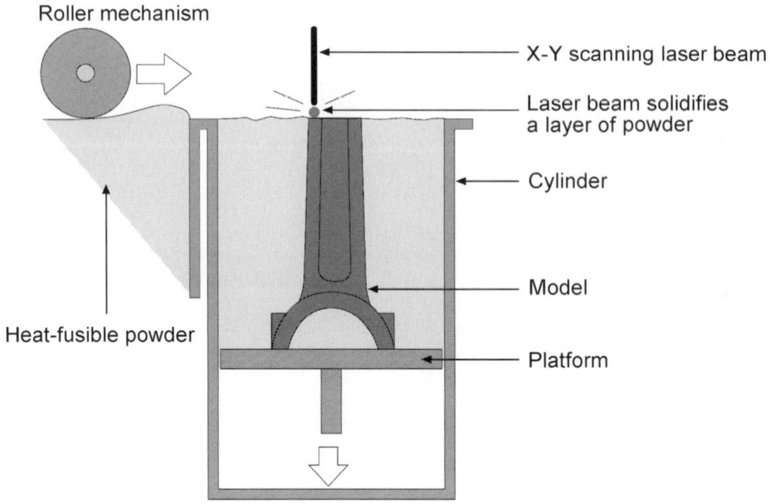

Figure 4.2 Selective laser sintering.

cross-section on the surface of the powder the localised heating causes the powder particles to melt and fuse together. The object is lowered into the build chamber and a roller mechanism deposits a layer of fresh powder on top of the previously melted cross-section. The process repeats until the object is built.

Because the object is surrounded by powder it is not normally necessary to build support structures with the SLS process. Upon completion the object is removed from the build chamber and the loose powder removed by careful brushing. Because the surface of the object has been in contact with powder it can be somewhat rough and porous. The 'shredding' of incompletely fused powder particles can be a problem prevented by dipping, painting or infusing the surface with epoxy or polyurethane. Probably the most common material used in the SLS process is nylon. This produces objects with good mechanical properties strong enough to produce good snap fits in objects such as harness buckles and other mechanical clipping devices.

4.6 Laminated object manufacturing (LOM)

The LOM process offered by the Helysis Company, California, builds objects by laminating together layers of laser-cut adhesive paper cross-sections (Figure 4.3). The sheet of heat-sensitive adhesive-coated paper is fed from a pay-off roll and bonded to the proceeding sheet by a pressure roller. Excess material outside the boundary of the cross-section is cut into small squares to facilitate part

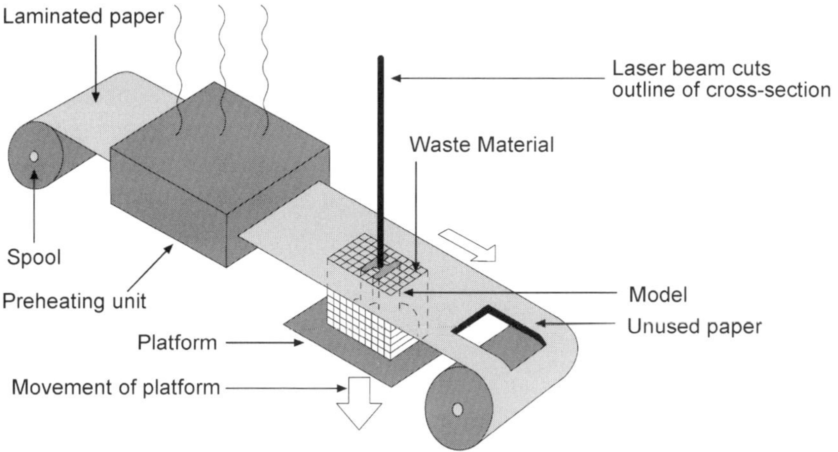

Figure 4.3 Laminated object manufacturing.

removal. The cross-section of each layer is cut with a carbon dioxide laser. When finished the object remains embedded in a rectangular block which acts as support material. The object is released by removing the excess material which has been cut up into small squares. To preserve the laminated paper object and reduce its tendency to absorb moisture, which can weaken the object and cause unwanted Z axis swelling, it can be coated with a suitable sealant such as wax, epoxy or polyurethane.

4.7 Fused deposition modelling (FDM)

This is a non-laser process and is offered by Stratsys, Minnesota. The FDM process builds its objects in thermoplastic materials such as wax, ABS, elastomers and a medically approved grade of ABS. It does this by depositing a very fine continuous molten bead of material from a small pen-sized vertical extruder (Figure 4.4). The extruder, fed with a monofilament via in-feed rollers from a spool, is moved in the XY axis like a printer plotter. The build platform is lowered to the Z axis to allow successive layers to be deposited one on top of the other. For large and complex parts a twin extruder system is used, with one extruder supplying the build material and the other a weaker support material.

Temperate control of the extruder and the build chamber is critical in order to ensure adequate bonding of the layers. The most common build material is ABS which lends itself to easy finishing by solvent wiping, spray painting or polishing. As for injection moulded ABS, fused deposition ABS objects can be glued or ultrasonically welded.

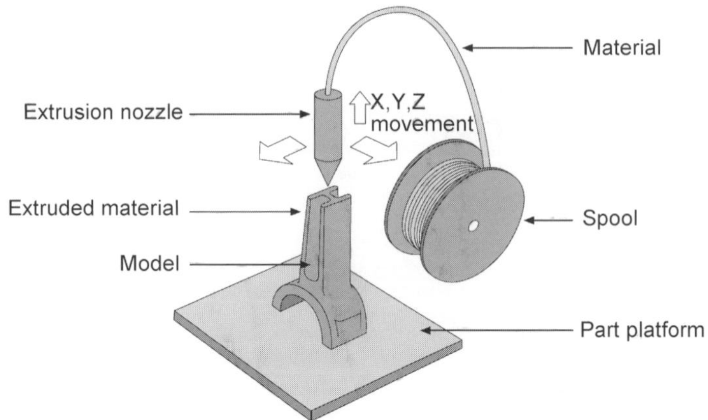

Figure 4.4 Fused deposition modelling.

4.8 Advantages of prototypes

The availability of a functional prototype at an early stage in a product's life cycle is a great advantage. It enables much useful information to be generated and this helps to refine the product's features, leading to the development of superior products and often reducing time to market. A prototype can be used:

- to validate the design from a functional and aesthetic point of view,
- for test marketing and focus groups,
- to demonstrate proof of concept to internal and external clients,
- to give something to the toolmaker when requesting tooling quotes,
- to enable secondary packaging to be designed and tried, and
- to enable change parts for filling lines to be made and tested before production bottles or other packaging are available.

4.9 Vacuum casting

A stage in between the rapid prototype and making production parts in the material of choice is vacuum casting (not to be confused with vacuum forming), using room temperature vulcanising (RTV) silicone tools and RTV polyurethane casting resins (Figure 4.5).

4.9.1 RTV silicone tool preparation

In the vacuum casting process, a pattern, typically made by one of the rapid prototyping systems, is used in conjunction with an RTV silicone rubber to

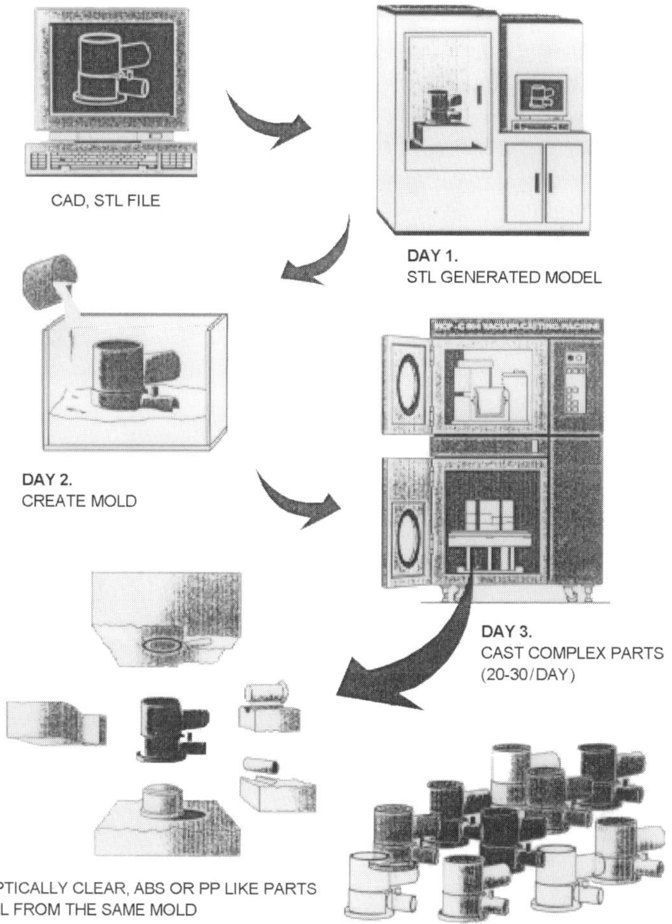

Figure 4.5 Vacuum casting.

produce a 'soft' tool. From this tool, several replicas are cast in a polyurethane whose properties approximate those of the material of choice from which the part will eventually be made. The main advantages of this process are speed—a tool can be made in a day—and low cost.

The basis of the vacuum casting system is the master pattern. It can be made of a material such as plastic, wood, metal or plaster or stereolithography (SL). Even temperature-sensitive models made from wax can be used without taking special precautions. A silicone rubber mould is cast from the master pattern.

Silicone rubbers are very viscous materials and, during the mixing together of the base rubber and catalyst, much air is entrapped. This must be removed by vacuum degassing since any air bubbles in the mould, especially those which may form on the surface of the pattern, will spoil the mould and render it unsuitable for further use.

Vacuum degassing is carried out by placing the mixed catalysed RTV silicone into a vacuum chamber evacuated to a pressure of 10–20 mbar for a period not exceeding 5 min. It is important to allow for an expansion of four to six times the original mixed volume during vacuum degassing. After the primary degassing, the silicone rubber is poured carefully around and over the model that has been previously prepared and mounted in a suitable casting box. Any deep closed (i.e. upside down) re-entrant features in the model may need venting by drilling small (<1 mm diameter) holes in the highest point of the feature or by pre-filling the feature with silicone rubber so as to avoid further air entrapment. To remove any air bubbles which may have been accidentally introduced while pouring the silicone rubber over the pattern, the casting box and its contents are replaced in the vacuum chamber for a secondary degassing. To ensure complete and controlled curing, the cast silicone rubber mould is placed in an air-circulating oven for 4–6 h at 40°C. When curing has been completed, the mould is removed from the casting box and the model released by cutting through the block of silicone rubber, to the previously delineated parting line, with a scalpel.

It is possible to cast short run production thermoforming tools using a similar technique. In this case an open or single-sided silicone casting is taken from a rapid prototype object. The silicone cast is placed feature face up in a suitable casting box and an aluminium filled epoxy or polyurethane is cast under vacuum on to the silicone. This method is very quick and produces fully dense durable thermoforming tools.

Some discussions have recently canvassed the idea of using rapid prototyping methods to produce thermoforming tools directly. This certainly can be done but tool life is expected be quite short and cycle times quite long because of the poor thermal conductivity of the rapid prototyping materials. By using a silicone intermediate casting, the rapid prototype object is preserved and if further thermoforming tools are required, either to replace damaged ones or simply to increase output, this is easily done.

4.9.2 Vacuum casting RTV polyurethane

To cast polyurethane replicas of the rapid prototype pattern, the mould halves are treated with a suitable mould release before being reassembled. The mould is usually held together with 50 mm wide adhesive tape; care is taken not to distort the mould when applying the tape.

Because the silicone rubber moulds are 'soft', they distort easily under pressure and so the polyurethanes are cast into them simply by gravity. To ensure complete filling, with no air being trapped during the casting, this operation is carried out in a suitable vacuum chamber.

Curing of the cast polyurethanes can take between 15 and 45 min, depending on formulation. With several moulds in operation, the process is quite productive. Depending on complexity, features and type of material being cast, it is possible to produce up to 10–20 castings from one silicone tool before the surface becomes brittle and starts to break up.

4.10 Use of rapid prototyping in packaging

Some examples of the use of rapid prototyping alone or combined with RTV silicone tooling and vacuum casting in the packaging industry are as follows:

Bottles. As block models using concept modelling; as rapid prototypes, usually by stereolithography but also by FDM; as vacuum castings: in this case it is common to make only a half bottle, split along the natural tool parting line, using a rapid prototyping method, then use this as a pattern for an RTV silicone rubber tool. Vacuum cast halves are glued together to make completed bottles.

One alternative is to make a complete bottle using a rapid prototyping method, cast an RTV silicone rubber tool from it and then produce replicas by rotational moulding with an RTV polyurethane. Another alternative is to create the bottle mould using a rapid prototype method and then use this to produce replicas by rotational moulding, again with RTV polyurethane. Both methods work.

Ice cream containers (up to 2 l) and lids. In this case the lid was vacuum cast in a polyurethane with similar flexibility to that of low density polyethylene and the base was cast in polyurethane with similar flexibility to that of polypropylene copolymer.

Other applications. Yoghurt cups, screw tops, aerosol can tops, meat trays, dispensing closures and egg trays are some of the many other applications. In the egg tray example two iterations were necessary.

The first design did not allow the eggs to sit low enough in the tray. This meant that when six trays were stacked up together they were too high to fit into existing packaging machinery. The redesign, which took only a few weeks to prototype and vacuum cast, proved satisfactory.

It was also possible to produce six additional castings for testing in a newly designed washing machine. The toolmaker was also provided with a vacuum cast tray as well as the NC tool path data. This helped speed up the tool making process by reducing the need for discussion and interpretation of drawings.

4.11 Keltool inserts

4.11.1 Introduction

The use of rapid prototyping has enabled product development and R & D engineers to reduce product development time cycles substantially. Typically, what used to take several months and involve considerable risk can now be achieved in a matter of weeks and with much more certainty.

By the use of silicone rubber tools and vacuum cast polyurethane, it is possible to replicate the rapid prototype master pattern and produce typically 10 to 20 parts in the space of a few days. Much useful information and data can be obtained using these methods, but eventually a production tool needs to be made to produce parts in an economic volume in the thermoplastic material of choice. With constant pressure on manufacturers to reduce the time to market and 'Do it once, do it right', the need for rapid production tooling becomes ever more important.

4.11.2 History of Keltool inserts

3D Keltool is a system for producing accurate, durable metal inserts using a casting, sintering, infiltrating process. Like any casting process, it uses a pattern and until the advent of rapid prototyping, and in particular stereolithography, Keltool was of interest but not a mainstream insert production method. The process was invented and developed in the mid-1970s by the 3M Company and called 'Tartan Tooling'; it was originally used to produce copper tungsten electrodes for spark eroding.

3M divested their interests in 1990, selling the technology to some of their previous employees, who renamed the process 'Keltool'. Keltool was acquired by 3D Systems—the makers of stereolithography apparatus—in 1996 and, yet again, renamed 3D Keltool.

4.11.3 Technology of Keltool inserts

3D Keltool inserts are tool steel inserts created from stereolithographic master patterns in a rapid prototyping time frame. A powdered metal/ceramic sintering technology is used to create tooling inserts rapidly for plastic injection moulding and die casting.

The process starts with an accurate properly finished pattern; like other casting processes, Keltool cannot improve on the quality of the pattern. From this pattern, a silicone rubber transfer mould is created. The metal powder, ceramic and binder mix is then cast into this transfer mould and allowed to set. Because the 'green' form is fragile, undercuts cannot be tolerated. The 'green' form, when demoulded from the transfer mould, is then put through the following stages:

debinding, sintering and finally infiltrating with a copper alloy. This gives a higher conductivity which shortens cycle time.

4.11.4 Properties

The end result is a tool insert with the approximate composition of 70% steel, 30% copper alloy and a minor amount of tungsten carbide. These inserts are produced with a hardness of 28–30 Rockwell C, but with suitable heat-treatment can achieve 40–44 Rockwell C. Due to the presence of the tungsten carbide, these inserts are quite suitable for sliding cores and have similar abrasion resistance to steel of 52–54 Rockwell C. The inserts machine similarly to H13.

All the usual metal working techniques, such as welding, grinding, photo-etching, drilling, tapping, texturing, polishing, electro-discharge machining (EDM), heat-treating and spark eroding, can be used on Keltool inserts. Due to process limitations, the optimum size for a Keltool insert should not exceed $200 \times 150 \times 125$ mm, but work is in progress to extend these boundaries. In many instances, several Keltool inserts have been successfully press fitted together to produce larger parts.

Unlike many traditional sintering processes, the Keltool shrinkage is low—only 0.6%—predictable and uniform. Shrinkage tends to be linear and isotropic. Accuracy is $\pm 0.2\%$, and flatness 0.001 inch per inch.

Because the process starts with a master pattern and transfer mould, the manufacture of multiple identical core and cavity sets is readily achieved. The usual turn-around time from receiving the master pattern to shipping the first metal insert set is 8 to 10 days.

The use of Keltool inserts offers the product design/product developer the possibility of short development times. It offers the toolmaker the possibility of making more tools in a shorter time and, because the inserts contain copper, Keltool inserts offer the moulder a decreased cycle time of up to 25%.

In summary, 3D Keltool is a well-established, commercially proven technology. It provides production quality metal tooling inserts in a rapid prototyping time frame. These inserts have the additional benefit of reducing the injection moulding cycle.

4.12 Conclusion

Speed to market is particularly important in the consumer goods industries. In the past, packaging development has frequently been accused of being the bottleneck in bringing new products to market quickly. With the developments in CAD, allied to the range of prototyping technologies available, it is now possible to meet the most demanding targets. Samples can be made available quickly for technical and consumer testing, eliminating many problems previously encountered in moving to production.

5 Design and decoration formats

M. Shickle

5.1 The role of design in determining packaging form and decoration

Walk down any supermarket aisle and you are faced with a wall of brands and product messages. Take a look at the variety of shapes, colours, typefaces and finishes, all of them designed to grab the consumer's attention for just a fraction of a second. Add to this the fact that consumers know what they are doing when they are shopping. They want value for money. They are not fooled by mere cosmetic changes to a package. They do not decide to buy a product simply because of packaging gimmicks that amuse or entertain without providing any extra benefit.

Given these demanding performance criteria, how does the packaging team enhance the pack and significantly boost the brand? How do the packaging manager and marketer come up with innovations to boost recognition and increase sales for a flagging product? Most of all, how does the development team step outside the confines of consumer understanding and find a new way of defining the product and brand messages? How do consumers rationalise their needs with what is on offer? This can be especially difficult when products replicate each other within a set of rules for forms and materials governing the category.

These category rules form an unwritten code for consumers to manage the selection process. Pack shape, material, surface texture, type of closure, decoration technique and graphic layout are the basic elements. They all create subtle responses in the consumer's mind that build a picture of the product and how it will satisfy the consumer's needs and desires. The challenge for packaging designers and marketers is how to exploit pack shape, decoration and materials to reinforce consumer perception and deliver added value beyond cost.

This chapter outlines some of the processes and models used at PSD to create distinctive and efficient packaging. They should not be seen as prescriptive or essential tools for a successful design programme. However, working with large, multidisciplined and sometimes geographically remote project teams, they provide a focus for shared creative decision making.

5.2 Managing pack requirements

The growth of modern supermarkets and use of plastics in packaging have been mutually linked. Today's supermarkets rely on efficient supply chains with packs that sell themselves on shelf and help consumers to get them home easily. Before the early 1970s plastic packaging lacked the technology to 'push' new ways of delivering the product into the home and the retail chain did not create the demand 'pull' from consumers.

Today, we have been taught to operate within an innovation culture. Any new material or process is interrogated to find an application from a retail and consumer perspective. Thermoplastics have also changed the way we think about solving a packaging challenge. This is because plastic conversion processes are varied—inject, blow, extrude and numerous hybrids—and we have more 'options' to apply to problems and to create innovative solutions. Just think of the number of ways of containing and dispensing a liquid using plastic. We will see later in this section, however, that there are rules and frameworks governing when and where to use pack forms, materials and decoration to define a product category and product positioning. Baked beans could feasibly be contained in a transparent laminated bottle or pouch but consumers will not relate to such a familiar product in unfamiliar clothes.

Great packaging design happens when designers combine as many elements of the functional performance and visual form of the pack as possible. What extra dispensing features can be added to a closure for no extra cost? How do we make a bottle easier to handle while reducing its weight? How can we exploit a decoration technique such as in-mould labelling or shrink sleeving to provide a physical pack function? None of these challenges, however, can make any sense without the context of the brief to define consumer and commercial requirements. Contrary to most opinions, the tighter and more demanding the brief, the better. Any design team enjoys finding ingenious ways to balance the interrelated goals of a packaging brief. Conflicts inevitably occur between various elements of the total pack mix. A closure, neck or handle required by consumers to dispense the product may be in conflict with a requirement to reduce bottle weight. The need for a larger label area on a plastic bottle may be in conflict with pallet utilisation, handle size, or fit on a filling line. Most pack material and decoration issues can be easily rationalised on the basis of polymer and product compatibility. Product actives or flavours can be easily lost through or absorbed by the walls of the container. Consumer handling requirements, opening features or extended shelf-life will drive another set of selection criteria. Every project is different and there is no set formula for listing priorities—these can only be qualified through consumer research and common sense. However, a good starting point for the packaging team to rationalise the requirements of the pack brief is to use an element from the Quality Function Deployment (QFD) matrix [1].

The feature 'correlation' matrix (Figure 5.1) allows comparison of consumer, performance and manufacturing requirements to highlight conflicts and compatibility. To create the matrix, key issues from the development brief are listed and prioritised. Setting an order at the start helps with the interpretation later. The example shown was used to build a better understanding of the project

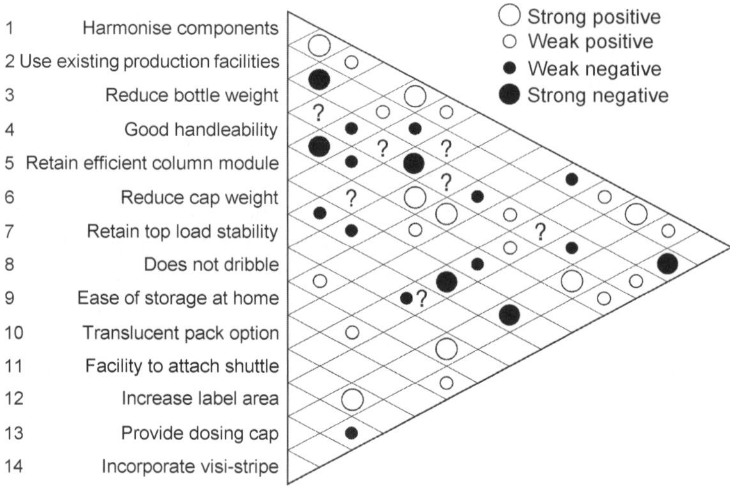

Figure 5.1 Feature 'correlation' matrix.

brief for a liquid detergent bottle and cap. Here, 14 packaging requirements are systematically compared with each other by following the diagonal rows until they meet. The team discussion required to score the outcome of each comparison is an effective catalyst for any packaging program. Generally, the process of team negotiation is more useful for creating a shared understanding of the design brief than a numerical outcome.

5.3 The material selection process

As a starting point we need to consider the product to be contained, its processing conditions, required shelf-life, distribution requirements, point of sale, usage and then disposal. Every functional requirement creates a filter for material selection. Although selection may seem like a mechanical checklist, compromises between cost, performance and fit with consumer needs will always be made. Design should be used as an exploratory process to extract the selection issues by visualising the potential functional benefits and visual impact of alternative materials. The key to success is the management of these compromises through knowledge of material, process and consumer insights.

PET has made large inroads in the beverage packaging market. Its immense strength, relatively high barrier performance and superb clarity have made it the ideal choice for the drinks market. Improved barrier performance is enabling PET to gain ground in the single-use beer bottle market. The properties of PET have also made it an ideal choice for the toiletry and personal care market. The superb clarity linked to new plastic labelling technology (the no-label look) make this an attractive package.

In markets where clarity or high barrier performance is less critical, such as the chilled cabinet sectors and for flavoured milk, the plastic sleeve decoration technology has opened up a market for simpler materials and standard structural packaging, such as HDPE for bottles with a full heat shrink plastic sleeve.

In the household and cleaner market many bottles are being made in HDPE with handles or off-centre necks, all designed to add functional benefits for the consumer and to characterise the brand at point of sale. Typically HDPE will be extrusion blow-moulded and as such, can have handles, or be multilayer to enhance the barrier or to include closed loop scrap material made up as an inner layer. As lightweighting has thinned the sidewalls it has become important to include a middle layer in a dark material in order not to see the product through the sidewall.

Rectangular thermoformed tubs dominate the yellow fats market. Toughened polystyrene and polypropylene are good materials that allow the design team to develop new tub forms and lids. More recently, to elevate the brand presence, injection moulding has been used to create premium-looking non-rectangular tubs designed for table-top presentation, through surface embossing and in-mould labels and even the use of basket-textured shrink sleeves for a rustic homemade appearance.

As food processing technology improves, new opportunities open up for plastics packaging. In the flexible plastics market the designer and packaging technologist will have the opportunity to select various laminate structures either for barrier, surface finish, tearing and integrating reclose features. In recent years flexible pouch packs have become a real alternative to conventional containers. They offer good tactile and visual cues to the consumer but are generally disliked by retailers for stacking and shelf utilisation. Owing to improved food processing technologies and the ever-increasing desires of marketers to stretch product positioning and margins, we are seeing cat food packed in pouches in place of metal cans. Here the designer can do more with the graphics and, as the technology progresses, we will see more shaped pouches made to reflect the end-use market or the brand expression.

Having narrowed the field down in terms of the choice of plastic to suit the product and consumer demand, the next consideration is the technology required to make the package. In many cases the plastic moulding technologies lend themselves to good levels of design flexibility. However, the designer and packaging technologist must consider the implications of annual product volumes and the product life-cycle to allow for capital investment, depreciation and amortisation.

Extrusion blow-moulding is relatively low cost for new moulds and so the commercial team can afford to relaunch products more frequently. As closures are injection or compression moulded, new tooling costs can be very high, and it is likely that a standard closure will be selected. Multilayer extrusion bottles are also relatively easy to develop and so this would not be a restriction.

Extrusion stretch and injection blow-moulded containers are still relatively easy to retool, although the preform and stretching process will make this more complex than a straight extrusion blow-moulded container. Injection blow-moulded containers require relatively expensive tooling so new designs may be limited to body changes only using standard performs and neck finishes. Here, selecting the right supplier will be important; one with a good portfolio of preforms and not too costly because of a constant changing over of moulding lines.

Flexible laminate pouches offer easy size modification for basic rectangular forms. Although pouch-shaping technology exists it has not yet gained high market share, primarily because of the cost comparison against lightweight bottles and sleeves. Nevertheless, there have been great advances in combining pouring spouts and reseal features for all pre-made and form-fill-seal pouches.

5.4 Brands, product communication and the impact on packaging shape

Consumers relate to products and brands in many ways. At a basic level, a pack and the product it contains must satisfy functional needs, but of crucial importance is the way the pack communicates its value and builds a relationship with consumers. For example, in the haircare market, premium brand shampoos align themselves with ‘professional’ high performance products used in hair salons. These bottles tend to have sharp, hard forms when compared to most other commercially available shampoo bottles that have been created to support rich and moisturising messages (Figure 5.2).

At the start of any packaging design programme it is useful to map the potential sensory messages that a pack form and decoration can deliver. These maps can have multiple axes to explore consumer perception of product shape, manufacturing complexity or competitive positioning. Figure 5.3 shows a simplified map used to model the perception of pack shape against manufacturing capabilities for a culinary two-part ‘cook-in sauce’. The resultant pack balanced strong wholesome food cues with experimental exotic cuisine.

To understand better how to create the ideal balance between the functional needs of product formulation, manufacturing capabilities, supply chain, and consumer brand proposition, the PSD Brand Hierachy© model [2] breaks down and orders brand characteristics. By characterising the brand or product in this way we can generate links to order three-dimensional functional packaging and two-dimensional surface graphics. Experience has shown that this is a

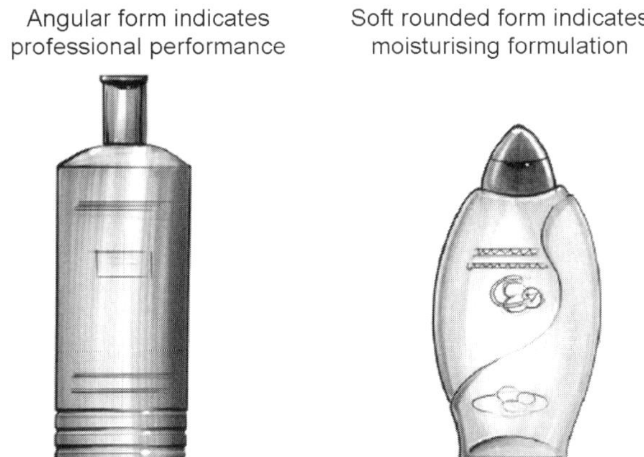

Figure 5.2 Shampoo bottles of different shapes are used to communicate a difference in the products.

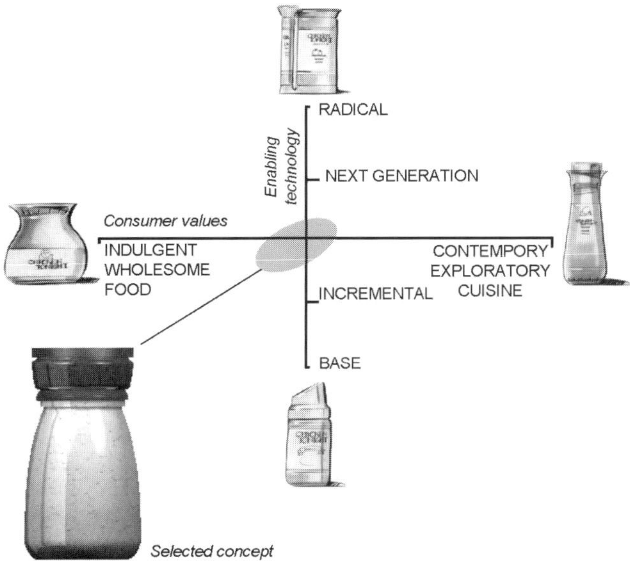

Figure 5.3 Model for perception of pack shape against manufacturing capability.

useful technique for exploring a packaging brief and driving consensus between packaging technology and consumer expertise.

In 1950s America, Abraham Maslow looked at the needs of modern society and gave labels to the various stages in individual development. He believed that our primary needs are physiological—food and drink. Having satisfied these basic needs, we require safety and security. This is followed by interest in

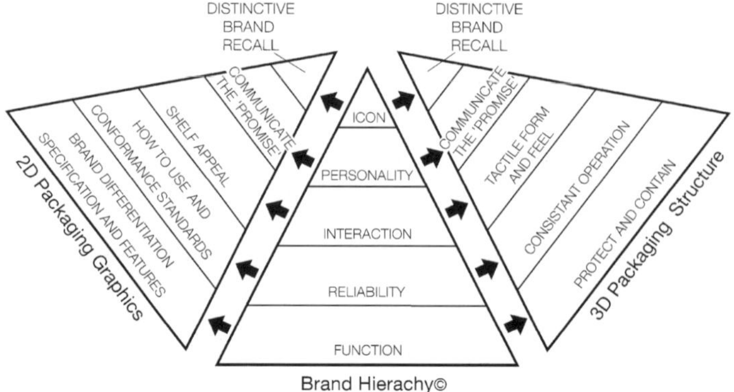

Figure 5.4 The PSD Brand Hierachy© model.

our social conditions, then a need for self-esteem and finally we might reach the fully developed human condition which Maslow called self-actualisation.

This simple but rather effective hierarchy has provided the framework for a Brand Hierarchy© (Figure 5.4). Working on the assumption that consumer brands are driven by the satisfaction of needs, a corresponding set of values has been defined which relate directly to the pack structure, graphics and formulation. These are:

1. *Functional need.* The product must do something useful and ideally offers...
2. *Reliability*, which allows a degree of...
3. *Consumer interaction*, taking us closer to the product's...
4. *Personality*, which ultimately results in...
5. *Iconic status.*

At the centre of Figure 5.4, the Brand Hierarchy© lists attributes of the product and brand perception. These values are then transposed for two-dimensional pack graphics and three-dimensional pack structure. The project team, including all packaging, marketing, logistic and trade functions, can then prioritise and associate specific pack issues driven out of the design brief. As with the QFD correlation matrix detailed at the start of this chapter, the process of negotiating to create a shared understanding of how the product and brand can drive pack perception is extremely important.

5.5 Brands, products and packs never stand still

The selection of packaging materials and decoration techniques is driven by category rules. However, innovations in materials, manufacturing process, stretching product capabilities and brand perception means these rules are never static. What was a discriminating or energising feature yesterday, is an essential basic feature of a product today. Toothpaste tube caps were once screw-on, screw-off components. Flip-top caps initially gave differentiation through convenience. Now consumers expect this feature as standard. Omission of a flip top in most markets today will be a barrier to purchase.

The model below shows the ACE matrix for consumer acceptance of a product or pack (Figure 5.5). Published in the *Harvard Business Review* by Ian C. MacMillan and Rita Gunther McGrath, May-June 1996, it builds on the 'Kano' model of quality, named after its inventor Dr. Noriaki Kano [2]. The ACE matrix is based on the premise that not all product or pack features are created equal; each cell of the matrix has a distinctive impact on the competitiveness of the product. The columns capture the level of energy that the product feature creates—whether customers regard it as a basic, discriminator, or an energising function. The rows reflect the sentiment that the attribute creates for the consumer.

5.5.1 Basic

A basic attribute is one that the target category has come to expect from all competitors. Basics are taken for granted: they rarely engender much loyalty or

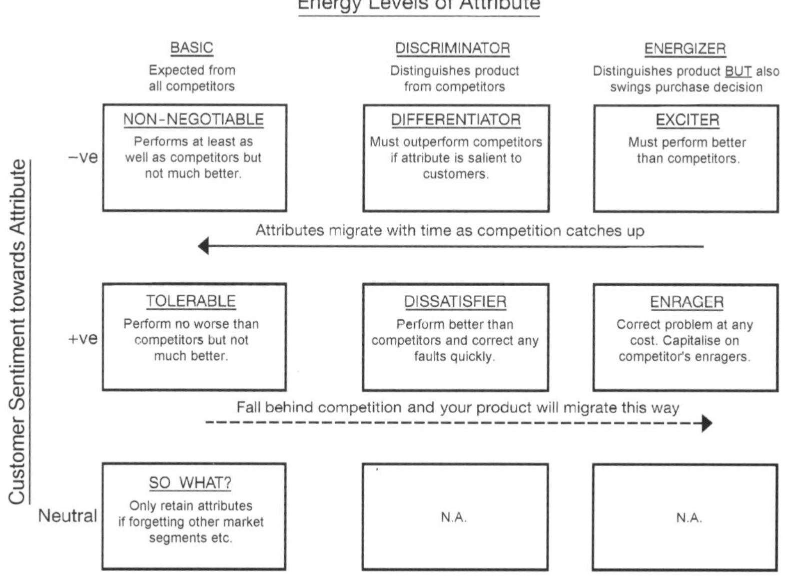

Figure 5.5 ACE matrix. *Harvard Business Review*, 1996.

antipathy. Basics are the attributes customers expect in any competitive offering. If customers feel positive about a basic attribute, it is called a non-negotiable because such an attribute is essential for a product to participate in the market, for example, beverage can ring-pulls. If customers feel negative about a basic, it is tolerable as long as the product performs no worse than its competitors—and as long as customers believe that the benefits of the offering transcend the inconveniences and that there would be the same inconveniences if they bought a competitive product.

5.5.2 Discriminator

A discriminator is an attribute that distinguishes a product from its competitors. If customers feel positive about a discriminator the attribute is a differentiator, for example, a shower gel hang-up hook. If customers feel negative about a decriminator, the attribute fits the dissatisfier cell of the matrix, and it is imperative for the company to ensure that the product performs no worse than competitors' products. The problem with dissatisfiers is that they can seriously erode loyalty and push customers into the arms of the competition.

5.5.3 Energiser

An energiser attribute is so powerful that it not only distinguishes a product from others but it often becomes the basis on which a purchase decision is made, for example, the flip open dispenser pack for mini POLO mints (Nestlé).

5.6 Summary

Good packaging is not solely about creating a fit between the conversion of polymers and the contained product. Producing great packaging is about much more. Consumers do not see the technology when they shop in the supermarket. They buy the pack on the basis of their needs and emotions aroused by the product and brand under which it is sold. The challenge for any packaging team is to build packs that key directly into the consumer's beliefs and values that give rise to selection and purchase.

References

1. Quality Functional Deployment and Kano's model are concisely summarised in *Product Design*, Mike Baxter, 1995, Chapman & Hall, London.
2. Brand Hierachy© is copyright of PSD associates Limited International design consultancy, telephone +44 (0)20 8614 7300.

6 Development of flexible plastics packaging

J. Dixon

6.1 Introduction

This chapter examines the criteria used to develop converted flexible packaging. Packaging materials that have been made from products such as paper, film and aluminium foil and converted by processes such as printing, coating, laminating, slitting or bag making are included. Development of the base materials is not considered. The converter can, and should, influence such development, by telling the manufacturer what performance is required. However, converted materials are usually custom-made; base materials are standard products. The development process is different, with different time scales and resources available. Similarly, the development of packaging machinery or pack shape, which is essentially defined by the machinery, is not discussed. Again, the converter can, and should, input into such development but it is a different process.

Anyone who wishes to develop flexible packaging has, at any given time, a fixed palette of materials from which to choose, in order to meet the needs of his customers, the packers. Different materials have different properties. For instance, LDPE film is tough and tear-resistant; oriented polyester film is heat-resistant; aluminium foil has excellent barrier properties. A summary is provided in Section 6.3. Sometimes a single layer of a material will meet a user's needs; at other times, different materials must be combined to provide the required properties. It should be noted that most other converters will have access to the same palette. To give the customer the differentiated packaging he requires, the successful converter must often be quicker and more skilled in selecting and combining the right materials.

The needs of customers must be clearly established for a development to be successful. This is easier said than done, especially when the product to be packed, or the packing process, is new.

Most converted packaging development is evolutionary. Brand new products to be packed or brand new packaging processes are quite rare. Most development is in response to a change in the customer's needs for an existing product or to a need to pack a new but closely-related product. The change in needs relates to:

- cost reduction
- shelf-life enhancement
- new packing machines and
- improved presentation

or a host of other factors which will be described below. The majority of present day packs owe their existence to a long history of step-by-step changes, often seemingly very minor and often stretching back over several decades. The history of the development of such packs would be interesting but inappropriate in a volume of this type. Instead, the criteria for the design and development of a packaging material are considered in detail. Often such criteria are not examined as a whole; only those that have changed are examined, thereby prompting a particular development. It may be useful to re-examine these criteria more often to prevent over-specifying or under-specifying.

The following section includes a table summarising some basic properties of common raw materials that make up the palette (Table 6.1). This gives a general guide to performance. Examples of some well known packs are considered and criteria which define their packaging needs are detailed to show how a particular structure meets these needs. These are only examples, mostly taken from the UK food industry. Other solutions are possible, and in some cases desirable. Finally a few comments are made on the design process itself.

6.2 Design criteria

6.2.1 Primary design criteria

These are the properties required for a flexible pack in order for it to be functional.

6.2.1.1 Containment

Packs have to contain a product to provide a unit of sale. To do this, the packaging material must have a certain strength. The type of strength depends on the pack style and the product packed. For example:

- tensile strength to prevent snapping or splitting under steady load
- impact strength to prevent bursting when packs are dropped
- tear strength to prevent ripping open, e.g. by abrasion
- puncture resistance against sharp objects and
- resistance to the chemical nature of a product, e.g. essential oils in spices

Such different types of strength are measured in different ways and will not necessarily coexist in a material; indeed, some are mutually incompatible. It is also necessary to have a means of closing or sealing the material; on the principle that a chain is only as strong as its weakest link, the seals need to be as ‘strong’ as the material (unless deliberately weakened to provide an easy opening feature).

6.2.1.2 Protection

The great majority of flexible packs do not merely contain a product; they also provide protection. At the most basic, this is against dirt and contamination. Other protection can be against the following factors.

- *Moisture pick up or loss.* This can be a prime cause of product loss of quality. The water vapour transmission rate (WVTR) gives a measure of the material's barrier properties against moisture.
- *Gas.* Oxygen can cause rancidity and other problems. The oxygen transmission rate (OTR) is often used as a measure of gas barrier. Permeability to carbon dioxide or nitrogen can also be important.
- *Odour.* It is important to keep intended flavours and scents in the product and to keep external contaminants out of the pack. The difference in chemical make-up of various odours makes it difficult to measure or predict odour barrier.
- *Light.* This can initiate oxidation reactions and loss of colour.
- *Mechanical damage.* With flexible packaging, such protection is provided by a combination of pack shape with material stiffness or rigidity. Entrapped air can provide cushioning.

6.2.1.3 Presentation

Most packs, especially those for retail sale, are required to make the product attractive to the consumer. Inks, coatings and materials can be chosen to give the desired combination of colour, gloss or matt finish, metallic effects, product visibility, etc. Shelf appeal can be enhanced by material properties such as rigidity—to make a pack stand up on the shelf—and resistance to creasing—to prevent it looking 'old' and 'tired'.

6.2.1.4 Information

The pack will usually be the medium through which the consumer is informed of the contents and how the product should be used. In some aspects, such as weight declaration, the pack is functioning as a legal document.

6.2.2 Secondary design criteria

While not absolutely required for pack functionality, these criteria must be satisfied for the pack to be a practical proposition.

6.2.2.1 Cost in use

The direct cost of the flexible materials is obviously important. A full cost evaluation should also measure the effect of different pack styles and specifications on the secondary and tertiary packaging material costs, on the packing line and distribution costs and on the waste levels through the whole process.

6.2.2.2 *Machineability*

A multitude of questions must be answered if materials are to form functional packs on high speed automatic packaging machines. Examples are listed below.

- *Sealing.* Will constant heat jaws, impulse seals or cold sealing be used? What quality of seal is required? Is sealing inside to inside (B:B), outside to outside (A:A) or inside to outside (B:A)?
- *Slip.* What slip or coefficient of friction (CoF), material to material or material to metal, is needed?
- *Stiffness.* Is rigidity in the material needed for efficient operation? Is dead-fold required to form the pack?

6.2.2.3 *Consumer needs*

How can the pack be opened easily? Is it 'single shot' or does it need to be reclosed?

6.2.2.4 *Environmental impact*

This needs to consider the full life of the product and some relevant issues are listed below.

- *Packaging production.* What emissions are generated by the manufacture of base materials and conversion of the material?
- *Disposal.* Can the material be recycled? What weight of material needs to be disposed of? How does one balance material minimisation with recyclability?
- *Product safety.* Are the packaging materials suitable for contact with the foodstuff or other product under the conditions of storage and use (e.g. temperature), without transferring anything that could be a health hazard or affect the product quality?

6.3 Basic properties of raw materials for flexible packaging

Table 6.1 summarises, in very simple terms, some key performance characteristics of common raw materials used for converted flexible packaging. It cannot give the whole picture because, for example, there are very many grades of low density polyethylene (LDPE) or oriented polypropylene (OPP) whose properties of barrier, strength and cost will vary considerably. The table is only intended as a guide to illustrate why one material might be chosen in place of another.

Table 6.1 Key performance characteristics of flexible packaging raw materials

Material	Barrier properties: Moisture transmission rate (37°C, 90%RH)	Barrier properties: Oxygen transmission rate (23°C, 0%RH)	Strength: Tensile	Strength: Resistance to tear propagation	Rigidity	Heat sealability	Heat resistance (shrinkage/ softening)	Clarity	Opacity	Cost
25 µm LDPE (extrusion coated)	++ (15[a])	0 (~ 6000[b])	+	+++	+	+++++	+ (~ 100[c])	+ (if matt finish)	++[d]	+
25 µm LDPE film	++ (15[a])	0 (~ 3500[b])	+	+++++	+	+++++	+ (~ 100[c])	++	+[d]	++
25 µm HDPE	+++ (7[a])	0 (~ 2500[b])	++	+++ (can be 'splitty')	+++	+++	++ (~ 130[c])	+	+[d]	++
25 µm cast PP	++ (12[a])	0 (~ 3000[b])	++	+++++	++	++++	++ (~ 130[c])	++++	+[d]	+++
20 µm co-extruded OPP	+++ (7[a])	0 (1700[b])	++++	+	+++	+++	+++ (~ 145[c])	++++	++[d]	+
35 µm cavitated OPP	+++ (6[a])	0 (1500[b])	++++	+	+++	+++	+++ (~ 145[c])	0	+++	+++
35 µm PVdC coated cavitated OPP	+++ (5[a])	++ (25[b])	++++	+	++++	++++	+++ (~ 145[c])	0	+++	+++++
12 µm oriented PET	++ (40[a])	+ (110[b])	+++	+	+++	0	++++	++++	0	++
15 µm oPA	+ (300[a])	++ (30[b])	++++	++	++	0	+++	++++	0	++++
40 g/m² paper	No barrier	No barrier	+++	++	+++++	0	+++++	0	++ to +++	++
9 µm aluminium foil	+++++ (<0.1[a])	+++++ (<0.1[b])	+	+	++	0	+++++	0	+++++	+++
Uncoated cellulose film	0 (1000+[a])	++ (10[b])	++	+	++++	0	+++++	++++	0	+++++
20 µm metallised OPP	++++ (0.3 to 1[a])	++ (10 to 50[b])	++++	+	+++	+++	+++	0	++++	++
12 µm metallised PET	++++ (1[a])	++++ (<1[b])	+++	+	+++	0	++++	0	++++	+++
Oxide coated 20 µm OPP	++++ (0.5[a])	+++ (<5[b])	++++	+	+++	+++	+++	++++	0	++++
Oxide coated 12 µm PET	++++ (0.5[a])	++++ (<0.5[b])	+++	+	+++	0	++++	++++	0	++++

Abbreviations: LDPE = low density polyethylene film; HDPE = high density polyethylene film; PP = polypropylene film; OPP = oriented polypropylene film; PVdC = polyvinylidene chloride; PET = oriented polyester film; oPA = oriented polyamide (nylon) film.

[a]units are g/m²/day.

[b]units are cc/m²/day.

[c]units are °C.

[d]if pigmented.

6.4 Case studies

6.4.1 Frozen vegetable bag

Deep frozen peas, beans and other free-flowing products are packed on vertical form-fill-seal machines in 'pillow pack' format. Pack weights can vary from 100 g to 5 kg. Traditionally, packs have been displayed flat in freezer 'chests'; more recently, eye level display cabinets have become popular. Table 6.2 summarises key design criteria for frozen vegetable bags.

The most common format (Figure 6.1) is a surface-printed white opaque modified polyethylene (PE) film—for example, ethylene vinyl acetate (EVA) copolymers, linear low density polyethylene (LLDPE) blends, metallocenes—to give the desired strength. Thickness is generally between 50 and 75 μm depending on pack weight.

The modified PEs are the only materials with the required deep freeze strength characteristics. As single webs, they are quite economical. They have the disadvantage of requiring 'impulse' sealing on the packaging machine, increasing both capital cost and maintenance costs compared to constant heat sealing. However, any attempt to laminate a more heat-resistant material to the outside to accommodate constant heat sealing impairs the deep freeze properties. The same reasoning would argue against using laminates to improve presentation (e.g. stiffness, gloss). However, structures such as OPP/PE and polyester (PET)/metallised polyethylene (met PE) are, in fact, being used increasingly

Table 6.2 Key design criteria for frozen vegetable bags

Criteria	Requirements	How satisfied
Strength	Resistance to puncture (from sharp, hard frozen vegetables) and to drop impact at −30°C.	Film with good elongation and high yield and ultimate tensile strengths at low temperatures.
Protection	Moderate barrier to moisture to prevent dehydration.	PE based films generally adequate.
Presentation	Freezer chest display options limited; good product illustration required.	Tone print of product (often process flexo) on white opaque film. Ink adhesion to resist high humidity.
Cost	Price-sensitive products requiring low cost packaging and high machine efficiencies	Surface printed monofilm. Lap seals on VFFS machine to save material usage.
Machineability	B:B and B:A seals made with 'impulse' jaws.	Low melt flow index (MFI) resin for strong seals. PE corona discharge treated for printing but stripe left untreated in lap seal area for better seal quality.
	Slip controlled for running over forming collar and down filling tube without tracking.	Slip additives in PE and ink system.

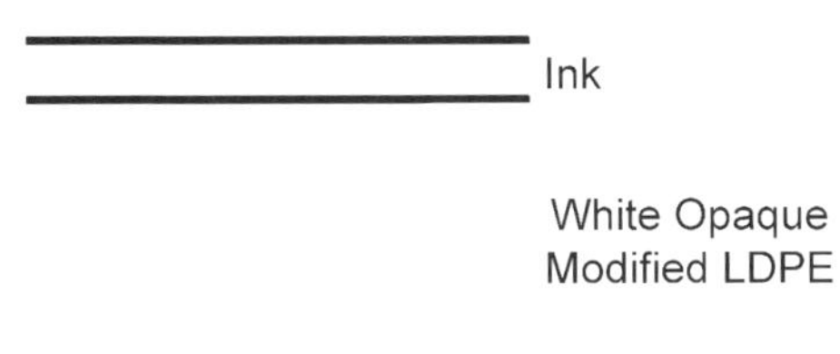

Figure 6.1 Ink/modified PE for frozen vegetables.

for higher value-added products, but at a cost in terms of both money and deep freeze performance.

Few pack designs pay much attention to the ease of opening by the consumer, nor to how the pack should be reclosed when half empty. It is probably thought that there is little wrong with using scissors and that the consumer's own wire tie works as well as an expensive proprietary reclose system, such as a zipper. Apart from solvent emission if such inks are used for printing, the manufacture of these packs causes little adverse environmental impact. The modified PEs are completely recyclable. The permitted monomers and additives for the films are well defined so there are few food contact issues.

6.4.2 Breakfast cereal liner

Most free-flowing breakfast cereals are packed in bag-in-box format. Pack size ranges from 30 g to 1 kg. The cereal is bagged on vertical form-fill-seal machines, typically running at 35 to 45 packs/min. Four such machines may feed an automatic cartoner. In a common end load design the bags lie flat and are pushed into the open end of a carton also lying flat. Table 6.3 summarises the key design criteria for breakfast cereal liners.

A co-extrusion of high density polyethylene (HDPE) and ionomer blend satisfies the design criteria (Figure 6.2). The thickness of the film can range from 35 to 80 μm depending on the barrier and strength required. Normally the HDPE forms around 85 to 90% of the structure.

The HDPE provides moisture barrier properties, strength, rigidity and a temperature differential compared with the inner ionomer blend, which provides low temperature sealing and peelability. A 'burst' peel mechanism has proved most effective (Figure 6.3). The inner seal layer welds to itself on sealing. Because it is only a thin layer, and because there is a weak, but controlled, interply bond between it and the HDPE, the seal can be peeled open by rupture of the sealant layer. Delamination from the HDPE propagates the peel.

This type of peel mechanism has a number of advantages. When the inner sealants weld together, they give a hermetic or near-hermetic seal, flowing round contamination. The peel initiation strength is essentially controlled by

Table 6.3 Key design criteria for breakfast cereal liners

Criteria	Requirements	How satisfied
Strength	Resistance to puncture by product.	Resins selected for toughness. In extreme cases, PP is used.
Sealing	Low temperature to achieve bagging speeds. Good hot tack to resist product fall onto hot seals. Ability to give good seals through product dust contamination.	Selection of resins with low seal threshold, high melt strength and good flow properties.
Protection	Moisture barrier of less than $5\,g/m^2/24\,h$ generally required; better for sugar-coated products or small packs.	HDPE in appropriate thickness meets most barrier needs.
Rigidity	Rigidity in the bag aids the cartoning process.	HDPE adequate.
Cost	Barrier properties per £ spend are key, provided other design criteria are met.	Low cost production method: co-extrusion of multilayer film.
Machineability	Constant heat seal jaws used to achieve seal integrity and peelability. Thermal differential between outside and inside surfaces needed to achieve machine speeds. Good hot tack to resist product fall. Control of CoF on both surfaces for forming collar, filling tube and cartoning process. Avoidance of static build-up.	Resins selected to provide a higher melting outer surface (HDPE at 130°C) and a lower melting inner (ionomer blend at about 90°C). Ionomer provides excellent hot tack. Careful dosage of additives to control slip and static.
Consumer needs	Peelable seals needed for easy opening of bag without tearing so that unused product is kept fresh. Bag reclosed by folding or with mechanical device, e.g. clothes peg.	Peelable ionomer blend. Film can be formulated for improved dead-fold.

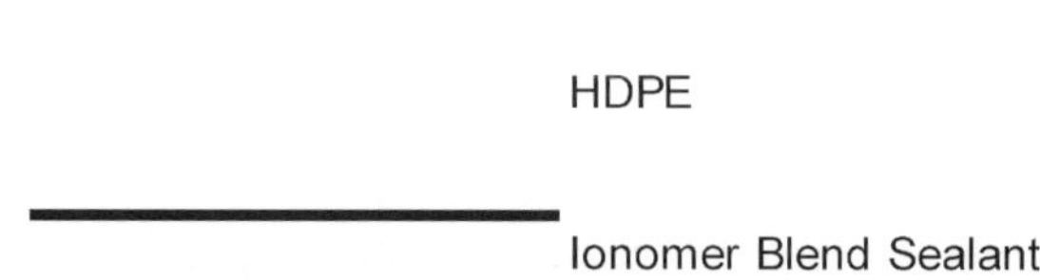

Figure 6.2 HDPE/ionomer co-extrusion for breakfast cereals.

the thickness and tensile strength of the sealant layer (Figure 6.4). It can be set at a level high enough to ensure seal integrity during cartoning and transport. The seal propagation is controlled by the interply bond and can be set lower so as to facilitate opening.

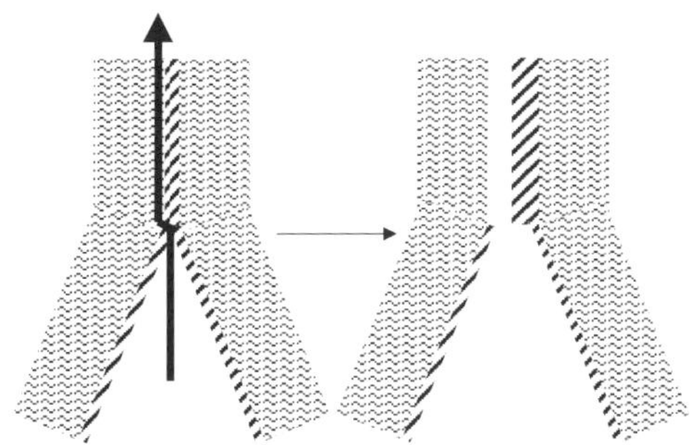

Figure 6.3 Burst peel mechanism.

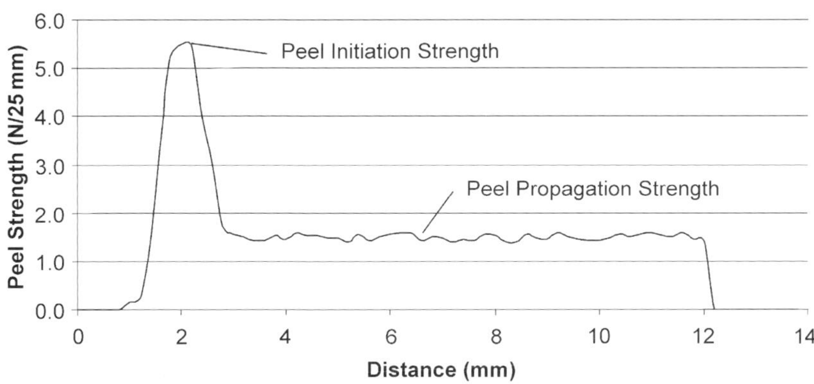

Figure 6.4 Initiation and propagation forces for a burst peel seal.

These co-extrusions result in very low environmental impact. There are no significant emissions during manufacture and both process and post-consumer waste are fully recyclable.

6.4.3 Tobacco carton overwrap

'Flip top' cartons for cigarettes are overwrapped in film. A sheet of film is cut and wrapped tightly around the carton with a lap seal along one edge and 'grocer folds' at each end. The packing processes for these products are both highly automated and very fast. Although there is generally a further collation pack to provide a unit of 200 cigarettes, for example, these are broken down in most sales outlets so that the overwrapped '20s' or '10s' carton is the display unit. Table 6.4 shows the key design criteria for tobacco carton overwraps.

Co-extruded OPP has replaced coated cellulose film as the material of choice for this application. Thicknesses are generally 15–25 μm. If printed, the ink is

Table 6.4 Key design criteria for tobacco carton overwraps

Criteria	Requirements	How satisfied
Protection	Moisture barrier needed to prevent tobacco drying out.	Use of OPP.
Presentation	High transparency and gloss.	Use of OPP.
Machineability	A:A, A:B and B:B seals. Tight slip and static control.	Co-extruded OPP with copolymer on both sides with specially formulated slip and antistatic additives.
Consumer needs	Easy open.	Tear tape.

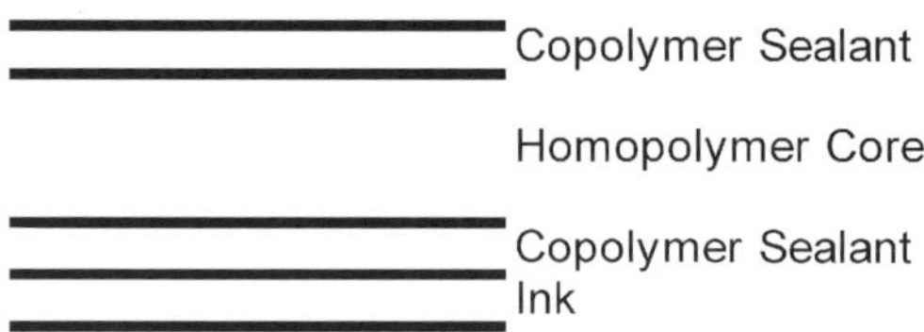

Figure 6.5 OPP/ink tobacco carton overwrap.

applied on the underside (Figure 6.5). The homopolymer core provides low haze (for transparency), good moisture barrier properties, stiffness, mechanical strength and heat resistance (shrinkage at around 145°C). The thin copolymer skins will start sealing at around 115°C; this gives a sufficient 'window' for efficient operation. The same functionality can be provided by coatings—for example, polyvinylidene chloride (PVdC) or acrylic—but, since applying these involves an extra production step, such coated films are more expensive.

Such films are frequently used unprinted since the carton provides all the decoration necessary. However, if additional information or promotional print is required, the overwrap can be printed, normally on the reverse so as to preserve gloss and minimise ink scuff. The slip of the plain film is carefully controlled for optimum machine performance. If inks are used, they must maintain CoF.

6.4.4 Biscuit roll wrap

The roll wrap is a common presentation for biscuits in the UK market. A stack of round biscuits is wrapped in a sheet of film. A fin seal is made along the length of the pack and sealed a second time down onto the side of the pack for improved appearance and seal quality (Figure 6.6). The material at the pack ends is folded and sealed against the end biscuits. Similar style packs are used for rectangular biscuits; the fold patterns will vary according to machine type and biscuit shape. Table 6.5 shows the key design criteria for biscuit roll wraps.

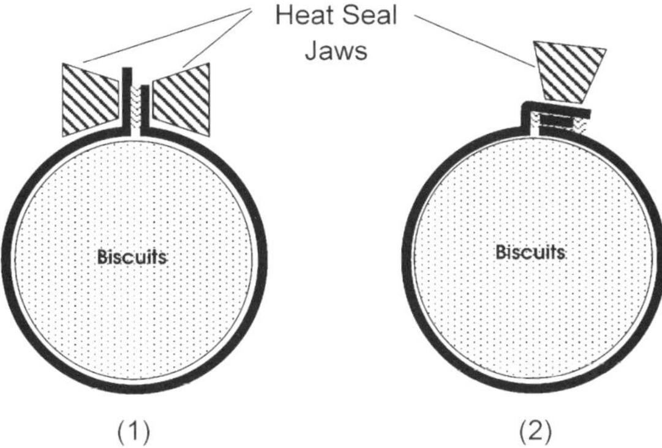

Figure 6.6 Long seal on biscuit roll wrap.

Table 6.5 Key design criteria for biscuit roll wraps

Criteria	Requirements	How satisfied
Protection	Moisture barrier.	Use of OPP.
	Oxygen and odour barrier.	Use of PVdC coated films.
	Mechanical damage.	Tight packs to hold biscuits firmly together.
Presentation	Usually illustration of biscuit plus strong branding through use of colour, gloss, matt, metallised effects. Colour in end seal area.	Quality tone print with tone illustration. Coating over ink to give sealability and gloss etc.
Machineability	A:A, A:B and B:B sealing.	Compatible sealants on both surfaces, e.g. PVdC or acrylic.
	Controlled CoF and jaw release.	Coating formulated with wax and other additives.

There are many variants possible but the following structure is very common for a roll wrap (Figure 6.7). The converter buys a cavitated OPP film which has been coated by the film manufacturer on one side with acrylic and on the other with PVdC. The converter prints (usually on the acrylic side) and overcoats with PVdC. Overall thickness is commonly 40 μm. The core of this structure is the OPP film which provides strength, rigidity and moisture barrier. Cavitated versions are most commonly used because:

(a) their inherent whiteness can often save the use of a white ink,
(b) their opacity prevents show through of the product, particularly important when packing chocolate-coated biscuits, and
(c) their better dead-fold characteristics aid formation of the pack and of good seals.

Solid OPPs are used when puncturing by, for example, a sugar-coated biscuit, would otherwise be a problem.

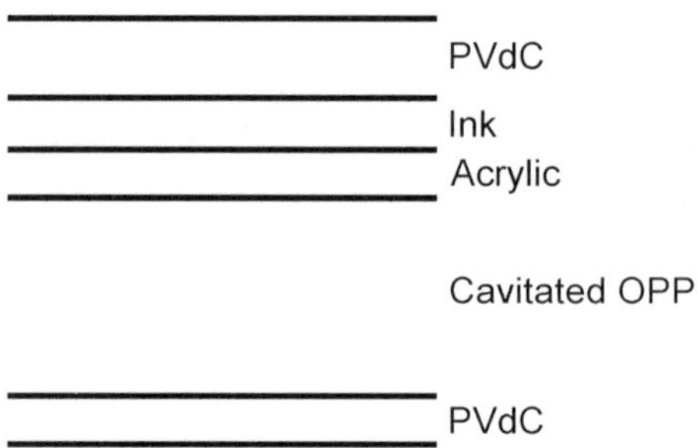

Figure 6.7 PVdC/ink/coated OPP for biscuit roll wrap.

For marketing reasons, the ends of the packs are generally required to be fully printed. If a co-extruded OPP were used, with copolymer sealing surfaces, either the ink would have to be capable of sealing to the copolymer or else it would have to be overcoated with a sealable lacquer. So far, no one has been able to develop an ink or lacquer that performs satisfactorily. In this structure, therefore, a coated OPP has been used. PVdC is the sealant on the inner surface. The ink is overcoated with a second layer of PVdC, normally 3 to 6 g/m^2 in weight. This PVdC coating performs a number of essential tasks:

- it seals both to itself and to the PVdC on the inside,
- it releases from the heat seal jaws (normally Teflon coated),
- it provides a glossy surface and protects the ink from scuff, and
- it gives a surface with a controlled slip for good machineability.

It should be noted that the top, converter-applied PVdC makes little contribution to the gas barrier of the structure. It is formulated specifically for sealing, jaw release, gloss and CoF. The gas barrier comes from the PVdC applied by the film manufacturer using a rather different formulation. In some countries chlorine-containing materials such as this are discouraged or not allowed because of the hydrogen chloride (and possibly dioxins) released when material is burnt in old or poorly controlled incinerators. In such cases, and if no gas barrier is required, acrylic coatings are used in place of PVdC. Such structures are in commercial use but rarely perform as well on the packaging machine as the PVdC coated version. Roll wraps are not easy to open, except with a sharp knife, and thus are usually fitted with tear tapes.

6.4.5 Potato crisp bag

The well known crisp bag is made on vertical form-fill-seal machines, usually with a lap seal. Pack size ranges from 25 to 150 g or more; only the smaller 'individual' packs are considered here. These are sold in a wide variety of retail outlets with a range of display formats; for example stacked on shelf, display racks, or straight from the corrugated outer. In any case, the pack must catch the eye of the consumer. Promotional activity is intense with most packs being special offer designs. Table 6.6 shows key design criteria for potato crisp bags.

Table 6.6 Key design criteria for potato crisp bags

Criteria	Requirements	How satisfied
Protection	Barrier to: (a) Moisture which makes product soggy (b) Light which initiates oxidation reactions (c) Oxygen which induces rancidity and off-flavours.	OPP provides a degree of moisture barrier but metallisation reduces permeability to below 1 $g/m^2/24\,h$. It also provides a light barrier and reduces OTR from around 1000 to less than 50 $cc/m^2/24\,h$.
Sealing	Ideally moisture and gas-tight seals, both A:B and B:B at high bagging speeds.	The copolymer surfaces of OPP heat seal—but not always hermetically.
Presentation	Eye-catching shelf appeal. Promotional activity.	Use of metal in the design. Flexo to allow frequent design changes.
Machineability	Controlled CoF	OPP additive package allowing slip transfer in reel.

The 'standard' potato crisp bag is made from an OPP/ink/adhesive metallised OPP (metOPP) laminate. The outer OPP is co-extruded with copolymer sealant and is usually 15 to 20 μm thick. It is reverse printed, usually flexo. This film is laminated, commonly with solventless polyurethane adhesive, to a metallised co-extruded OPP also 15 to 20 μm thick (Figure 6.8).

Until recently, small crisp bags were surface-printed, co-extruded OPP. The search for improved product quality led companies to adopt metallised films as

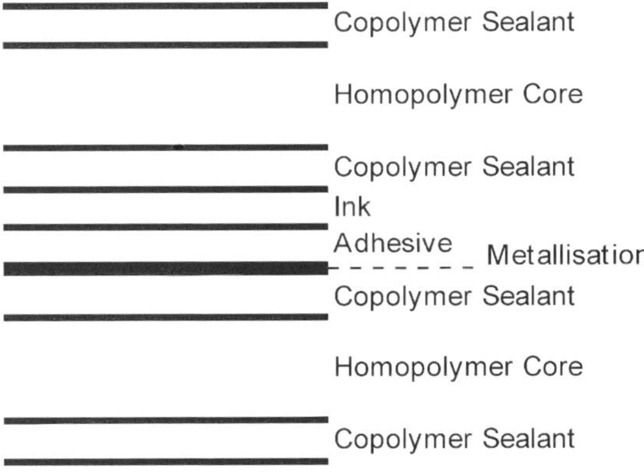

Figure 6.8 Potato crisp bag laminate.

a means of protecting product from the UV light which induces rancidity. A surface printed metOPP would, in theory, have done the job but a number of other factors were important, as listed below.

(a) The relatively fragile metal surface would be vulnerable to damage during the printing process and would thereafter only be protected by a thin coat of ink and lacquer. Further damage could occur during packing and distribution.
(b) Unless expensive stripe metallisation were used, lap seals would not be possible, necessitating the use of additional material (3.5 to 5%) and modification of forming shoulders of existing machines.
(c) Surface print on metal results in relatively poor gloss and whiteness. The current structure overcomes these problems. The metal surface has fewer rollers to pass over and hence opportunities for damage when run on a laminating machine. In the laminate, it is protected by a layer of adhesive and an OPP. Indeed some adhesives may even enhance the barrier properties by 'filling up' any holes in the metal.

The A surface of the outer OPP can lap seal to the B surface of the inner OPP. This A surface can also be controlled to give optimum CoF. To enhance metal adhesion, metOPPs have to be made migratory additive free, which results in a high CoF on their sealant surface. Film manufacturers have now developed additives in the A surface which will transfer to the B surface when they are in contact in reels of laminate. This effect lowers the B surface CoF sufficiently for efficient machine running.

The reverse print, of course, gives excellent gloss. The whiteness compared to surface print is also enhanced since the ink is lifted by the adhesive away from the greying effect of the metal. These improvements in presentation can also be enhanced by a greater or lesser use of the metal itself showing through unprinted areas or through transparent inks.

The metallised OPP offers an additional moisture and gas barrier, as well as protection against UV light. Most major crisp manufacturers are now gas flushing their packs so a gas barrier is needed to maintain the oxygen-free atmosphere. However, this additional barrier is worthless unless seals can be made reasonably hermetic. This is not easy; the copolymer layers are less than 2 μm thick. Seals may be contaminated with salt and product. At the junction of the fin seal with the cross seal, the thickness of the material increases from two to three layers and then back to two again. This implies a step of around 40 μm, which results in a channel along its edge. It is very demanding to ask 4 μm of copolymer to fill such a gap. Similar potential channels exist at the extreme ends of the cross seal. Despite improvements in copolymer sealing qualities and jaw design, these difficulties are not yet completely overcome. Lamination using PE instead of adhesive is a promising technique; the PE softens and flows in the heat seal jaws to enable the channels to be filled better.

6.4.6 *Confectionery countline flow wrap*

Chocolate countlines are almost all packed in flow wraps made on horizontal form-fill-seal machines. Normally they are high volume products. A minority are sold in multipacks but the majority are sold as single units from counter displays. These are strongly branded impulse purchases. Table 6.7 shows key design criteria for countline flow wraps.

Table 6.7 Key design criteria for countline flow wraps

Criteria	Requirements	How satisfied
Protection	For most of these products, the enrobed chocolate acts as a barrier for the filling.	OPP generally adequate.
Presentation	Shelf appeal and strong brand image.	Consistent colour and logos. Gloss finishes.
	No chocolate show through by 'wetting out' on film.	Use of cavitated OPP.
Machineability	High speed operation on horizontal form-fill-seal (HFFS) machines (up to 700 bars/min).	Coldseal as sealing medium.
	Low extensibility under tension to maintain pitch repeat.	Oriented film, e.g. OPP.

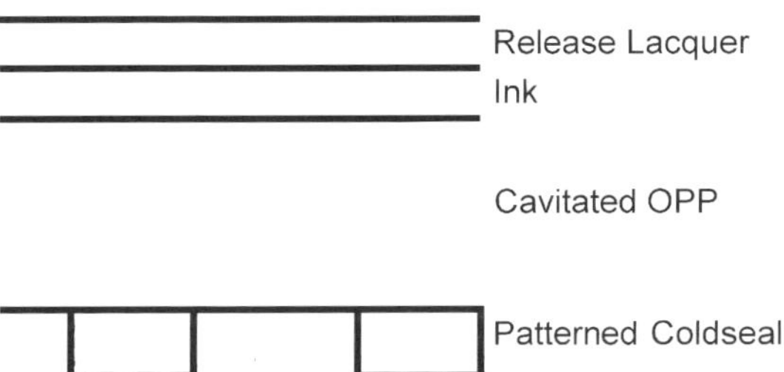

Figure 6.9 Confectionery flow wrap.

The most common confectionery flow wrap is a cavitated OPP, 35 to 50 μm thick, which is surface printed. Coldseal adhesive is applied to the underside, patterned in a 'window frame' so that it only appears in the areas of the film that form the seals (Figure 6.9). A release lacquer over the ink ensures that the coldseal does not block in the reel. With a few notable exceptions, the traditional countline pack styles of inner foil wrap with band or strap of paper and of die-fold wrap of foil/tissue or glassine have given way to the OPP based flow wrap. OPP is chosen for a number of reasons.

(a) Its barrier properties are adequate for the great majority of products. If better gas and flavour barrier is needed, it can be coated, for example with PVdC. If a better moisture barrier is needed, it can be metallised on the outer surface.

(b) High speed flow wrappers, operating between 45 and 120 m/min, exert considerable force on the material as they pull it over the forming box. OPP has sufficient tensile strength to resist elongation and consequent loss of register of the patterned coldseal.
(c) When chocolate comes into direct contact with a clear or white solid film, even if it is printed, it can show through as an unattractive dark shadow. Cavitated films, with their superior opacity, prevent this show through, hence the widespread use of cavitated OPP. Such films can also be made very white, eliminating the need to use white ink.
(d) Although cavitated OPP costs more than solid, it is still an economical material.

Heat sealable OPPs can, and do, run on countline flow wrappers. Modern grades can achieve speeds of up to 80 m/min. However, the use of coldseal is more common. Packing machines can be simpler with no jaw heaters and controllers and no cooling plates. This makes them cheaper to buy and to maintain. There is no need to lift off heaters when machines stop, or to adjust temperatures with changes in running speed. This is useful on fully automatic lines when packing machines may ramp up and down in speed to adjust to variable availability of product. Finally, high speed heat seal films are generally less widely available and more expensive. Thus coldseal is used to seal these packs; 3 to 4.5 g/m^2 of coldseal will seal to itself instantly under pressure alone, with no heat. Such seals will generally not be hermetic but they are quite adequate for this application.

Coldseals are generally pattern applied for a number of reasons, listed below.

(a) They have poor slip properties; their surface is somewhat tacky. If applied all over, they can tend to grip the product as it is pushed into the formed tube of material during the packing process. A coldseal-free area allows the product to slide against the OPP.
(b) Coldseals have a distinctive odour. Although they should always be suitable for direct food contact and regularly tested to ensure that they do not in any way taint the chocolate, it is still prudent to reduce risk by keeping them away from the product as much as possible.
(c) Why apply, and pay for, 100% coverage when only 25 to 40% coverage is needed to do the job?

A patterned coldseal does require that it is applied in register with the print and that, in turn, the pack is registered to the print. This is no disadvantage so far as pack appearance is concerned. It poses some extra challenges for the printer in terms of achieving this registration and maintaining a pitch repeat tolerance that is suitable for the packing machine, but this is seldom an issue on modern presses with a stable web of OPP.

The only other issue with coldseal is how to stop it sticking when that is not wanted, i.e. when it is wound under pressure in the reel. Coldseal will block to conventional inks and a number of other surfaces. Polyamide-based print-over lacquers possess release properties and will allow free unwinding of the material from the reel. Formulation and application of these release lacquers—or rather of the combined release lacquer/ink system, since they should always be considered together—is a key issue in the manufacture of such products. Not only must the system give good release, it must:

(a) give a controlled CoF. Too high a CoF will cause the web to screech when running over the forming shoulders or prevent running altogether. Too low a CoF can result in difficulties handling the product for collation or multi-packing;
(b) be capable of receiving an overprint for date coding, etc
(c) have good gloss and clarity to present the product.

The consumer can find these packs somewhat frustrating to open. The adoption of serrated knives allows a tear to be initiated from the pack ends but the tear is difficult to control. Cross direction or machine direction tear tapes are sometimes used to good effect. A number of other ingenious 'easy open' devices have been proposed but use has been limited by their cost and tendency to diminish machine efficiencies.

6.4.7 Modified atmosphere tray and lidding

Modified atmosphere packaging (MAP) is a technique for extending the shelf-life of fresh, normally chilled, foods. The atmosphere surrounding the product, typically fresh meat, fish, cooked meat or pasta, is a controlled mixture of gases designed to inhibit the spoilage process. The precise blend will vary from product to product. During storage, there will be a change in the mix due to respiration by the product and diffusion through the pack walls. To minimise this, it is helpful to have as large an air volume to product ratio as possible. One way of achieving this is by using three-dimensional semi-rigid trays with lids.

A common packing method uses thermoform-fill-seal machines. A semi-rigid base web is formed into a tray, filled with product and, after flushing with the correct gas mix, is lidded with a flexible film. The packs have a shelf-life measured in days or weeks rather than months, and are displayed in chill cabinets. Table 6.8 shows key design criteria for MAP trays and lidding.

There are a number of ways of meeting these design criteria. One typical structure is described here, which could be used, for example, for packing fresh meat. The tray material is thick (250–750 μm) ply of amorphous polyester (aPET) or polyvinyl chloride (PVC) laminated to an inner sealant of LDPE, normally 50 μm, often a peelable grade. The top web is oriented polyester film coated with clear oxide or PVdC for gas barrier and laminated to a 50 μm LDPE sealant. The latter is either formulated with 'anti-mist' additives or has an anti-mist coating on its sealant surface (Figure 6.10).

Table 6.8 Key design criteria for MAP trays and lidding

Criteria	Requirements	How satisfied
Rigidity	Base web must maintain tray shape.	Use of thick material with high modulus.
Thermoforming	Base web must thermoform.	Material with low softening point.
Protection	Moisture barrier to prevent product drying out. Gas barrier to maintain gas mix.	Easily satisfied by e.g. 50 μm LDPE. Either coatings (PVdC or oxide), or through bulk of base web materials or use of PE/EVOH co-extrusions.
Presentation	Product generally required to be visible without water droplets or mist on lid.	Clear top web with formulation/coating to increase inner surface tension.
Sealing	Gas-tight seals, even through product contamination.	PE used as sealing medium.
Consumer needs	Easy peel open. Reclose for some products, e.g. cooked meat.	PE formulation plus peel tab. Various proprietary designs available.

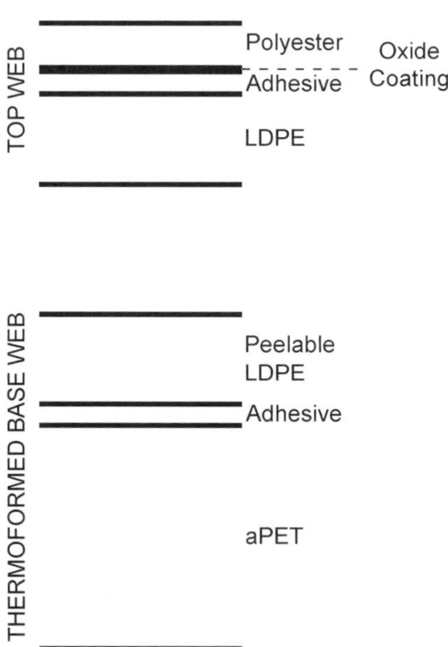

Figure 6.10 MAP tray and lidding.

For several years, PVC was the only material used as the structural element in the base web. It is easily thermoformable, rigid and offers a reasonable barrier to both gas and moisture. It can be made transparent or, if desired, opaque and coloured. Concern over the incineration of such chlorine-containing compounds has led to the substitution of aPET for PVC in some markets despite a cost premium and some increased difficulties in thermoforming and cutting. More recently, there has been a trend to the use of foamed polystyrene and polypropylene to provide opacity and rigidity. However, these polymers must be combined with ethylene vinyl alcohol (EVOH) layers to provide the required gas barrier.

The barrier properties of the top web were traditionally provided by a PVdC coated polyester. Again, concerns over chlorine content have led to replacement of the PVdC by other coatings, such as polyvinyl alcohol or oxide. The latter, either silicon oxide or aluminium oxide provide barrier by a very thin (less than 100 nm) layer of inorganic coating. Such coatings are applied in vacuum by a process similar to that of metallisation. They must be protected from mechanical damage by being laminated to another film.

Completely gas-tight seals are needed on MAP packs even though there is a real risk of the product contaminating the seal area while it is being filled into the tray. A 50 μm layer of LDPE is ideal for sealing round and through this contamination. Hence, a 50 μm film is laminated to both the base aPET and the top PET. The adhesive for the base must be chosen so as not to inhibit the thermoforming; it must 'flow' while still giving a good bond. The adhesive for the top must again give a good bond and be compatible with the oxide coating.

The polythenes are, of course, welded together by heat. OPP would be feasible as the base for the oxide coating but the greater heat resistance of PET allows for higher temperatures to be used and faster sealing speeds.

Such welded seals could only be opened with a knife or scissors so it is quite usual to design a peelable seal. This can be achieved by using a peelable resin blend for the sealant on the base layer. Putting the peelable layer here lessens the risk of 'stringing'—threads of delaminated PE forming on the inside edge of the peeled seal—and any diminution of clarity due to the peel blend will be more acceptable on the base.

A clear top web is essential for the consumer to be able to inspect the cut of meat, fish or other product. With standard PEs, moisture from the product will condense on the inside forming an opaque mist. To prevent this, the PE surface is modified either by additives in the resin or by a thin coating applied during or after the lamination process. The aim is to increase the surface tension such that the moisture wets out and water droplets, if formed, are large enough to see through.

6.4.8 *Soup sachets*

Dehydrated soups—and similar products such as dried sauces and chocolate drinks—are usually packed in a sachet format. These are made on horizontal sachet machines, with or without a bottom gusset. They may be displayed standing on end in cardboard display units. Since the dehydrated product is less than appetising to look at, the pack usually features a good illustration of

Table 6.9 Key design criteria for soup sachets

Criteria	Requirements	How satisfied
Containment	Resistance to puncture from particulates	LDPE sealant. In extreme cases ionomer or even additional ply of PET or oriented polyamide (oPA)
Protection	Hermetic seal even through powder contaminant.	PE (or ionomer) as sealant.
	Excellent barrier to moisture, gas and light all important.	Aluminium foil or metallised or oxide coated film.
Presentation	Rigidity for pack to 'stand up' on shelf. High quality graphics with good gloss.	Paper. Gravure print on coated paper (which also gives white background) + gloss lacquer.
Machineability	Rigidity for sachet to maintain shape in grippers during filling. Inside to outside temperature differential.	Paper. PE on inside, heat-resistant paper and lacquer on outside.
Consumer needs	Easy open.	Paper for easy heat initiation/propagation—enhanced with laser performance if needed.

the made-up soup and/or its ingredients. Alternatively, 'instant' soups are often packed four sachets to a carton. Their decoration is much simpler. Although the bulk of the product is a powder, some products also have quite large pieces of meat, vegetables or croutons. Table 6.9 shows key design criteria for soup sachets.

The classic sachet is based on a paper/PE/foil/primer/PE structure. This section will concentrate on the individual sachet which is displayed on the shelf. This will have its multicolour print protected by quite a thick layer (3 g/m^2) of high gloss lacquer; cartoned sachets have no lacquer. The paper will be 40 to 70 g/m^2 clay coated for display packs, and around 40 g/m^2 machine glazed bleached kraft (MGBK) for cartoned packs. The foil thickness will be no more than 8 μm, and often 7 μm or even 6.3 μm. The inside PE sealant will range from 12 to 35 g/m^2 depending on product type. The laminant PE is typically 12 g/m^2 (Figure 6.11).

The soup sachet is a classic case of a complex laminate where the individual layers play a variety of roles. The lacquer not only gives gloss, it also protects the ink from scuff and is highly heat-resistant, in some cases up to 240°C.

The clay-coated paper gives the structure bulk and rigidity and the larger the pack, the heavier the paper. Its inherent heat resistance allows the use of jaw temperatures up to 250°C, necessary on packing machines which can run at 400 sachets/min. The clay coating gives a smooth surface to print on and an inherent background whiteness. Finally, paper is easily torn for the opening of the sachet.

Aluminium foil gives an excellent barrier to moisture, gas and light. It is, however, easily cracked and pinholed, especially when used in very light gauges.

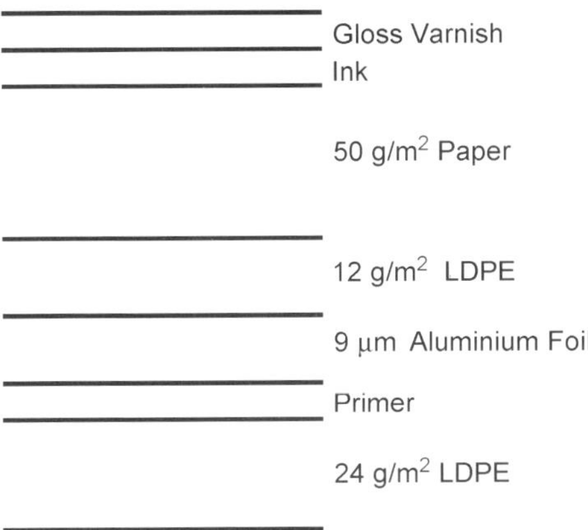

Figure 6.11 Paper/PE/foil/PE for soup sachets.

In addition to bonding the foil to the paper, the laminant PE also protects the foil from physical damage.

The PE on the other side also has this role in addition to its main one of closing the pack with B:B seals. The top seal is frequently contaminated with dust; PE can flow round such contamination to give hermetic seals. Some 'modified' PEs, such as ionomers, are even more effective at this. This laminate appears to be a complex structure and might be expected to be expensive. In fact, people have struggled to provide alternatives with the same functionality at lower cost because the laminate components (PE resin, paper and even lightweight aluminium foil) are relatively low cost materials. They are all combined in a single extrusion lamination and coating operation that can run at high speed. Bulk (i.e. stiffness for machineability and appearance) can be achieved at low cost by increasing paper and PE weights. To do the same with, for example, pre-made films, is more expensive.

6.4.9 Shelf-stable pouch for ready meals

These products have been common in Japan for many years but are only recently being seen in the European market in any numbers. A pre-made pouch, often of the stand up type, is filled with product (e.g. soup, sauce or prepared meat dish) and is sealed. The pouch and its contents are retorted at, for example, 121°C for 20 min, to sterilise the pouch and its contents. The pack has a shelf-life of up to a year. It is either packed into a carton or is displayed (usually in stand-up form) on the supermarket shelf. Many packs are designed to be reheated in the microwave by the consumer. Table 6.10 shows key design criteria for shelf-stable pouches.

There are a number of structures designed for this application; one suitable for a range of products and for use in stand-up format is described here.

Table 6.10 Key design criteria for shelf-stable pouches

Criteria	Requirements	How satisfied
Containment	High material and seal strength to resist drop and pressure tests.	Use of oPA in laminate. Strong lamination bonds and high yield strength sealant.
Protection	Moderate moisture barrier. Very high gas barrier.	Use of aluminium foil, oxide coated film or EVOH.
	Absolutely hermetic seals.	Choice of sealant in adequate thickness.
	Materials to withstand retorting without barrier loss.	PET, oPA, aluminium foil, oxide coatings and cast PP.
Presentation	High quality graphics.	Multicolour reverse printing.
	Visibility of some products.	All film structure.
	Stand up well.	Adequate thickness of sealant.
Consumer needs	Reheating with microwave.	No foil in structure.
Environmental	Food safety paramount.	Selection of components and testing under retort conditions. Pre-made pouched for more consistent seal quality.

A 12 µm oriented polyester film is coated under vacuum with a thin layer of oxide (e.g. silicon oxide, aluminium oxide or even a combination). The oxide surface is reverse printed and adhesive laminated to a 15 µm biaxially oriented polyamide film (oPA). This, in turn, is adhesive laminated to a cast polypropylene, whose gauge will depend on pack size; 80 µm would be typical for a 250 g pack, more for larger sizes (Figure 6.12).

Even if the product has been properly retorted and is effectively sterile, oxygen will still cause loss of product quality through enzyme-catalysed and other chemical reactions. The product will be degassed before packing and oxygen will be excluded from any headspace. The packaging material still has

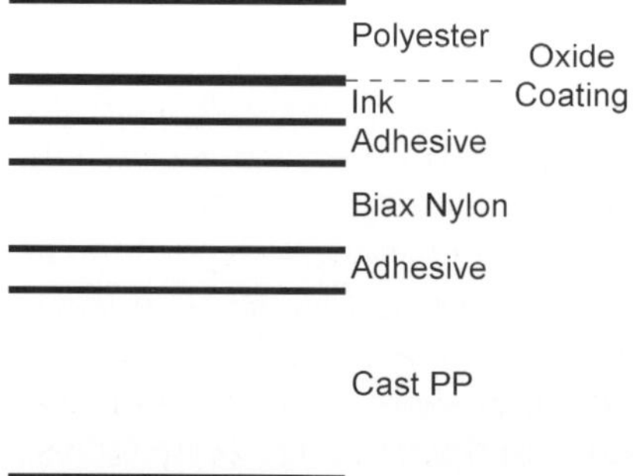

Figure 6.12 Retortable pouch.

to prevent oxygen entering. The use of aluminium foil is one solution but this is not an option if product visibility or pack microwaveability is required. EVOH is sometimes used but its barrier properties are much reduced under the high moisture conditions of retorting. There is a period while the EVOH is 'drying out' when the oxygen permeability may be unacceptably high. Oxide coated films are now available in glass clear grades, which are capable of giving laminated barriers of less than $2\,cc/m^2/day$ even after retorting. These are therefore becoming a cost-effective solution.

Such coatings are applied to oriented polyester that has high temperature resistance and a high tensile strength. This gives it stability in the vacuum coating process and during subsequent conversion. A 'stretchy' film may allow cracking of the oxide coating. Polyester's thermal resistance allows high temperatures to be applied during both pouch making and final sealing so as to ensure leak-free seams.

However, polyester's stability is combined with a low elongation at break and poor stress crack resistance. To give the pack strength and robustness, a tough oriented nylon is laminated to the oxide layer/ink. This operation also helps protect the oxide layer from mechanical damage, enhancing its real barrier performance.

The choice of sealant is very much a compromise. Cast polypropylene is the only real candidate. The resin grade and extrusion conditions must produce a film that:

(a) is heat sealable with sufficient flow to provide channel free seals but without excessive thinning (e.g. at the seal edges) so as to weaken them;
(b) maintains seal strength at retort temperatures so as to resist internal pack pressure in the retort;
(c) is tough (extensible) enough to withstand, for example, drop tests;
(d) is not embrittled, shrunk or discoloured by the product during retorting; and
(e) is rigid so as to give the pack stand-up qualities at lowest possible thickness.

Some of these requirements work against each other and a compromise has to be reached. In the future co-extrusion techniques will be valuable in achieving such a compromise.

The adhesives used have a great influence over the success or failure of this structure. These and all the other materials must, of course, be safe; that is, meet all the relevant FDA and European requirements for retortable packaging. The adhesives must bond the layers firmly yet be sufficiently flexible so as not to embrittle the structure, both before and after retorting. They must not detract from the clarity of the structure nor should they affect the barrier of the oxide coating.

6.5 The development process

6.5.1 In theory

A complete packaging system is seldom developed from scratch. However, it is still useful to look at what one should do in these circumstances since it raises issues that need addressing even in a more limited development.

The flow chart in Figure 6.13 outlines the main steps. It is, of course, vital to understand and agree the design criteria as detailed above. If these criteria are not yet all known—as will often be the case with new products—tests may need to be set up to establish design requirements. The next step is to consider 'what

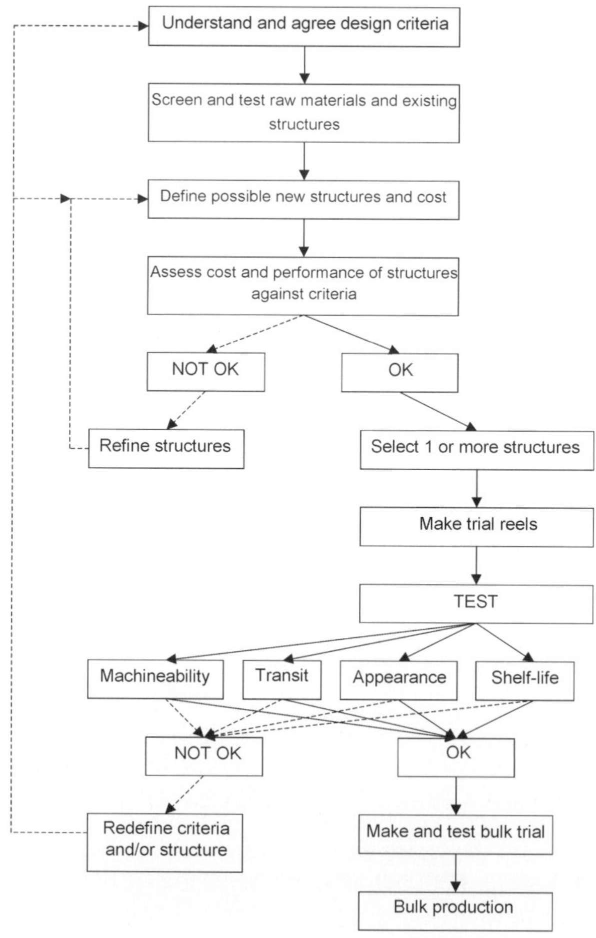

Figure 6.13 The development process.

might meet these requirements'. Often there are existing structures and materials whose properties are well known and can be compared with the requirements. In other cases, tests will have to be done. In the end, it is anticipated that one or more candidates can be costed, so that all parties can agree the economic viability of the solutions before incurring extra cost. If satisfactory, trial rolls of one or more specifications can be manufactured. If not, it will be necessary to define the system again.

The type of trial will depend on what questions remain to be answered. If one is absolutely certain that the structure will satisfy all the design requirements, there is, in theory, no need for a trial except to confirm viability of the pack. This is unlikely. The trial should be designed so as to address those issues over which there is least certainty. For example:

- prove pack shelf-life—unprinted web, packs perhaps handmade
- confirm pack sizing or machine performance— eyemark printed web
- approve pack appearance—fully printed web, perhaps done digitally.

If the trial fails in any way, it will probably be necessary to go back to the design stage or even to question the basic design criteria. If successful, however, it would be normal to scale-up production, perhaps supplying fully printed material where the previous trial had been unprinted or running on production plant as opposed to pilot plant. At all stages of this process, it is important that both the converted trials and the packs produced from them are thoroughly tested since such results will form the basis of performance specifications for ongoing bulk production.

6.5.2 *In practice*

In the more common 'evolutionary' type of development, this general process should still be followed but in an abbreviated form. Usually the design criteria and the extent to which the current structure satisfies them will be well known. The development will be focused on improving one particular aspect of performance, without affecting the satisfactory nature of other aspects. Much of the material and pack evaluation will concentrate on comparison of properties rather than achieving theoretical values. Thus one might seek to match or better the moisture barrier of an existing structure with a new material rather than going back to original data on food type and shelf-life. The prudent may still carry out shelf-life tests on the final packs but often effort will concentrate on comparing performance of new with old rather than looking at absolute figures.

6.5.3 *Summary*

The development of flexible packaging can range from the creation of brand new pack forms and material structures to the apparently minor—but still critical—amendment of a component. In all cases, it is important to have a good appreciation of a variety of material properties so as to understand how they can play a role in the final structure and how they will affect final pack performance. Given the wide variety of structures used, pack types and products packed, much of this knowledge is empirical; this does not lessen its value nor mean that one should not always follow a well-defined development process.

7 Plastics in active packaging

M.L. Rooney

7.1 Introduction

Successful packaging is achieved when the properties of the package match the requirements of the contents from packing until use. Packaged goods often have multiple requirements and, when considered with the limitations imposed by the packaging machinery, offer substantial challenges to existing packaging technology. An additional requirement is the need for many packs to perform roles in the use of the product. Such roles include those of the cooking vessel, dispenser or, more commonly, the advertising medium which promotes the product when the package is handled for the first time.

Since packaging also performs multiple roles in containment, protection, preservation and identification of packaged products it is possible to assemble a long list of desirable properties, many of which are mutually exclusive. The requirements of the product must therefore be prioritised and we find that there is less flexibility in packaging choice than might ideally be desired. Common examples are found where foods requiring barriers only to ultraviolet light and oxygen are packaged in foil laminates which are opaque to visible light and impermeable to water vapour.

An increasingly popular approach to meeting one or more of these needs is to develop active packaging which is designed to overcome specific limitations in existing passive packaging. Packaging is defined as 'active' when it performs some desired role other than to provide only an inert barrier to the external environment (Rooney, 1995).

Specific reasons for developing active packaging plastics are summarised in Table 7.1. These approaches to active packaging are elaborated in this chapter in the context of plastics as packaging media.

7.2 Overview of active packaging plastics

Plastics are used in several different packaging forms and in each of these forms appropriate methods of introduction of 'activity' have been developed or are the subject of research. Figure 7.1 summarises the forms of packaging in which plastics are commonly used and provides a reference point for an overview of appropriate methods of generating activity.

Table 7.1 Functions of active packaging plastics

Package activity	Examples of items affected
Removal of undesired component from the product or the package headspace	Oxygen, water, ethylene, odours, taints, carbon dioxide, heat
Addition of desired component to the product or package headspace	Carbon dioxide, water, antimicrobials, aromas, sulfur dioxide, heat, volatile corrosion inhibitors
Generating an antimicrobial surface	Microorganisms
Selective change to one or more physical properties of the packaging	Gas permeability, microwave absorption/reflection, porosity

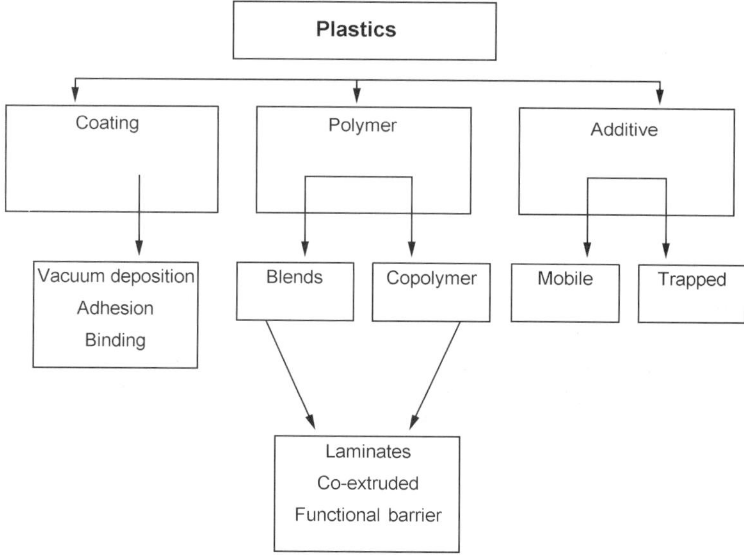

Figure 7.1 Forms of plastics used in active packaging.

The polymer itself constitutes the major component of packaging plastics and this polymer may in itself be active under certain circumstances. This polymer is frequently a copolymer or a blend of different polymers. This provides opportunities for generation of activity by selection of particular comonomers or by choice of the polymers in the blend. The plastic containing the active polymer may subsequently need to be combined with other materials in the form of laminates, co-extrusions or closures, or made into structures which include a functional barrier to one or more substance.

A simpler approach to achieving active packaging plastics is the inclusion of additives in commodity polymers during compounding of the material or during packaging material fabrication. These additives are either mobile or fixed depending upon whether or not they are needed to migrate into the package

Table 7.2 Effects achieved with polymers in active packaging

Effect	Role of polymer
− Oxygen	Reactant, matrix
− Ethylene	Matrix, solvent
+ Sulfur dioxide	Matrix
± Water vapour	Liner
± Carbon dioxide	Solvent
− Aldehydes	Reactant
− Thiols, amines	Matrix
+ VCI*	Solvent
+ Antioxidants	Solvent
+ Antimicrobials	Reactant, matrix
+ release	
− absorption	

*Volatile corrosion inhibitor

contents. Some additives need to be dissolved although insoluble powders are used where a cloudy or granular appearance is still acceptable.

Use may also be made of coating technology which is already highly developed in a range of forms. Simple application of active adhesive compositions has been proposed. Solvent or emulsion coating is also used and there have been some instances of research involving vacuum deposition. The most complex form of coating involves chemical binding of the active agent to the polymer surface (Halek and Garg, 1989).

The manner in which the polymers in some plastics have been used either commercially or in processes under development are summarised in Table 7.2. In some cases the same activity can be achieved by the plastics in multiple forms or compositions, as has been demonstrated commercially in the case of oxygen scavenging. In this case plastics have been involved either as matrices for dispersed solids, solvents for soluble additives or the reactive polymer or copolymer, either alone or as blends. These polymers have also been applied as a coating from a solvent or as a layer in a multilayer structure. In the case of humidity control cartons the plastics multilayer is being developed as a liner in a corrugated fibreboard package (Patterson and Joyce, 1993).

The commercialisation of active packaging has been occurring at different rates, depending upon whether physical or chemical processes are involved. Susceptors which convert microwave energy into heat, and films made porous by inclusion of particles or by controlled perforation have been used commercially for several years (Sacharow, 1995; Yam and Lee, 1995). The introduction of plastics based on chemical processes, such as oxygen scavenging in plastics films, have been marketed at an accelerating rate recently as a result of the discovery of chemical systems acceptable to regulatory authorities in some countries

(Anon., 1996a). The patent literature indicates the existence of a backlog of developments yet to be evaluated by regulatory authorities for food packaging.

This chapter surveys the roles played by plastics in the wide range of active packaging processes that are in use or under research or development. The underlying principles are discussed and some of the opportunities for their application are described by means of examples. Plastics formulations are considered as:

(a) media with functional additives,
(b) reactive polymers, and
(c) composite structures.

Each field of plastics formulation is discussed in terms of the opportunities for application.

7.3 Plastics with functional additives

7.3.1 Removal of undesired components

7.3.1.1 Oxygen scavengers

The removal of some of the oxygen from packages of sensitive foods and beverages is normally achieved by vacuum packaging or by nitrogen flushing. Commercial flushing typically leaves 0.5–2% oxygen so development of sachets containing oxygen scavenging materials has been particularly useful to gain much greater control of quality, particularly when residual oxygen left in a package is significant (Smith *et al.*, 1995). These technologies are not convenient to apply in all situations so a great deal of research is being directed towards oxygen scavenging plastic films.

The shelf-life of many foods is limited by uptake of as little as 1–200 mg/kg (Koros, 1990). The range of foods which could benefit from oxygen scavenging plastics, and the suitable forms in which the scavenger might be applied have been discussed (Rooney, 1995; Floros *et al.*, 1997). Oxygen scavenging films offer the prospect of being very versatile as their thickness can be varied to match the amount of oxygen to be removed.

The use of a scavenging plastic to provide a chemical oxygen barrier offers the opportunity to cheapen barrier packaging for relatively short shelf-life products such as wholesale or export units. This can be achieved by using a barrier plastic with inadequate barrier and upgrading its performance with a scavenger layer as in the first such laminates reported by Rooney and Holland (1979), or the more recent co-extrusions (Thomas, 1998).

The need to achieve oxygen removal from foods and other sensitive products at a rate faster than that of the degradation reactions or the growth rate of micro-organisms (or insects) means that only fast oxidation reactions of the polymer can be considered. Since premature reaction of the polymer with oxygen from the air must be avoided, there are approaches to accelerate oxidation reactions, usually involving some form of activation or triggering. The three most common methods of acceleration are water absorption, transition-metal catalysis and light excitation.

Oxygen scavenging plastics require the use of an outer passive barrier layer to oxygen ingress, or blended with a low permeability polymer in order to minimise the load on the chemical system. Polymers which provide suitable physical barriers are ethylene vinyl alcohol copolymer (EVOH) or a poly(vinylidene chloride)-coated layer.

7.3.1.2 Dissolved reagents

Very few oxygen scavenging plastics involving dissolved additives have been developed. However, reactions of such additives provide essential models upon which to base development of fully polymeric scavenging systems. A basis for determining the maximum rate of reaction of oxygen in a plastic is found in singlet oxygen chemistry (Rooney and Holland, 1979). Singlet oxygen is a highly reactive excited form of molecular oxygen readily generated within a plastic by photosensitisation. The rate of reaction can approach the diffusion limited value depending upon the reagents chosen.

If a polymer film containing a photosensitising dye and an electron-rich oxidisable compound, such as a furan derivative, is illuminated with UV, visible or near infra-red irradiation of appropriate wavelengths, oxygen is rapidly scavenged from within the plastic and is replaced by absorption from the surrounding air. The process commences with excitation of the dye by absorption of light, followed by rapid transfer of the excitation energy to oxygen dissolved in the plastic. The excited singlet oxygen thus formed diffuses through the polymer until it reacts with the acceptor, or becomes deactivated. The process continues until the oxygen or acceptor is consumed or until the source of light is removed. The scavenging of oxygen from the package headspace (for instance) is most rapid in plastics of high oxygen permeability, which allows oxygen molecules to come very close to the excited, immobile dye molecules during the triplet lifetime (10–1000 μs) of the latter. Once the light source is removed the process stops. Singlet oxygen can diffuse around 100 Å before being deactivated back to the ground state depending on the nature of the polymer matrix (Turro *et al.*, 1981).

Singlet oxygen chemistry has been used as an oxygen scavenging process where the polymer matrix is, for instance, cellulose acetate or ethyl cellulose (Rooney *et al.*, 1981; Rooney, 1982a). The sensitisers included erythrosin or *meso*-tetraphenylporphine and the acceptors were bis(furfurylidene) pentaerythritol and ascorbic acid. It was found that the permeability of the polymer film is an important determinant of scavenging rate with the ethyl cellulose being a better matrix for rapid scavenging than is cellulose acetate.

Ethyl cellulose has an oxygen permeability coefficient at 25°C ten times that of cellulose acetate. It was also found that the rate of oxygen scavenging by the film from a pouch is limited by light intensity initially but becomes diffusion limited when the oxygen partial pressure reaches low values (Rooney *et al.*, 1981).

Since illumination with white light of polymer films containing only one dye is wasteful of energy potentially capable of energising oxygen scavenging, the use of multiple dyes has recently been reported (Maloba *et al.*, 1996). It has been found that curcumin, a natural food colour and a poor photosensitiser, enhances the rate of oxygen scavenging by energy transfer to the eosine photosensitiser. The effectiveness of the film in suppressing rancidity in sunflower oil was found to be substantial at 23°C and 37°C.

The need for continuous illumination with visible light to remove headspace oxygen from a package has limited opportunities for application commercially. Hence, processes have been developed which rely on light only for activation of a variety of oxidation reactions of polymeric reagents as described later in this chapter.

7.3.1.3 Dispersed reagents

The most common approach to inclusion of active additives in plastics has been to disperse solid reagents, particularly those which require the presence of water to initiate or accelerate the reaction with oxygen. The water may dissolve the reagents or may plasticise the (polar) polymer sufficiently to allow a diffusion-limited reaction to proceed at an acceptable rate. This method of accelerating oxygen scavenging has been commercialised and constitutes the most popular approach to formulating crown closure liners for glass beer bottles (Teumac, 1995).

Amoco Chemicals (now BP Amoco) developed a master batch system containing iron and other components for use in polyolefin or PET structures. The products are termed AMOSORB™ and are water-activated, preferably by retorting. The first product incorporating this material is a closure liner termed TRI-SO_2B™. A patent application by Amoco Chemicals describes the use of iron in the presence of organic acids in polymer films (Venkateshwaren *et al.*, 1996) but the composition of the product AMOSORB™ is not revealed, being described only as consisting of substances GRAS in the USA. The absorbent layer will be separated from the food by a sealant layer at least 12.5 μm thick in film structures and 25 μm thick in multilayer laminate sheets. The product is not subject to premature reaction provided the masterbatches are stored at a relative humidity below 40% (Tsai, 1996).

Toyo Seikan Kaisha Ltd has produced Oxyguard™ trays with iron dispersed in sandwiched layers, but so far with only limited market penetration despite their simplicity of construction. The difficulty experienced in making rusting reactions occur in a hydrophobic polymer matrix is that the sealant polymer is

Table 7.3 Factors affecting oxygen scavenging by plastics

Effects of scavenging chemistry on physical properties
Access of co-reactant, such as water, to reagents in the plastic
Limitation of scavenging rate by oxygen permeability of the polymer
Appearance if solid particles are dispersed in the plastic
Premature reaction if the chemistry is not triggerable
Migration of any low molecular weight components or by-products

normally relatively impermeable to water vapour at ambient temperatures and so such systems are best applied to retort packages. Several factors have an effect on successful research outcomes in the field of oxygen scavenging plastics and these are summarised in Table 7.3.

Although several plastics have been formulated by blending of low molecular weight reagents into polymers, the most attractive alternative to dispersal of iron in plastics is organic reactions of polymers themselves. It appears that the rapid reaction rate achieved with high surface area iron powder in sachets is likely to be matched in plastics only by very rapid reactions, provided that the oxygen permeability of the plastic is not the limiting variable.

7.3.2 Absorbers of odours and food constituents

The concept of absorption of undesirable food constituents by packaging polymers is not new. Sulfur-resistant lacquers have been supplied commercially for many years in the tinplate canning of foods in which protein degradation resulted in the release of sulfur compounds from the food. These sulfur compounds cause the phenomenon of 'sulfur staining' on the tinplate and the lacquers containing zinc oxide react with these compounds before they can diffuse to the tinplate surface.

Early developments in odour-absorbent plastics occurred in Japan where there was seen to be a need to remove amine smells from fish which was stored in domestic refrigerators (Labuza and Breene, 1989). The amines formed in fish muscle degradation have potentially strong interactions with acidic compounds such as citric or other food acids. Hence, the earliest work involved incorporating such acids in heat-seal polymers like polyethylene and extruding them as layers in packaging.

Subsequently the ANICO company in Japan introduced pouches made from commodity plastics in which a ferrous salt and an organic acid such as citric or ascorbic acid is dispersed. This system is claimed to oxidise the amine or other oxidisable compounds which diffuse into the plastic.

There may well be a wide range of food constituents which can be removed by making use of specific interactions with selected packaging components or by chemical reaction with them. A fertile research field would seem to be open, especially with liquid foods since solubility and diffusion of food constituents in the packaging can be utilised so that the removal process is not limited to compounds with a significant vapour pressure at distribution temperatures.

An important example of removal of a food component by utilising diffusion is debittering of grapefruit juice. Bitterness in this juice is due to the presence of naringin and it has been shown that this compound can be hydrolysed to non-bitter products by the enzyme naringinase. Recent research has shown that immobilisation of the enzyme in cellulose triacetate film can cause hydrolysis of 60 to 80% of the naringin within 15 days at 7°C (Soares and Hotchkiss, 1998). The naringin is postulated as diffusing into the polymer in order to reach the enzyme. Several polymers were evaluated and the most effective was the cellulosic polymer, which is relatively permeable to large organic compounds. This effect shows the potential for the bitterness to be reduced to acceptable levels during commercial distribution rather than by use of additional processing equipment to bring the enzyme into contact with the juice.

Ethylene is often found undesirable in packages of respiring horticultural produce owing to its role in catalysing ripening and senescence of the produce. The gas is routinely removed during bulk storage by catalytic oxidation or by reaction with potassium permanganate. There have been many attempts to incorporate absorbents and reagents into plastic films used for carton liners without any process being free from drawbacks (Zagory, 1995). The main challenge has been to incorporate strong oxidants, such as permanganates, into polymers without damaging the latter.

An alternative approach utilises the Diels–Alder reaction of ethylene with the double bonds in tetrazines resulting in the binding of the ethylene and release of nitrogen (Holland, 1992). So far only small tetrazine derivatives have been used as solutions in polymer films such as PVC. This approach has been shown to remove headspace ethylene concentrations to the low parts per billion level at concentrations at which it is unable to cause petal loss in sensitive flowers like carnations. These films are self-indicating, being pink initially and bleached on reaction with ethylene, which occurs rapidly at concentrations that are physiologically important. Since only a low (ppm) tetrazine concentration is required in the packaging film, there is reason to be optimistic that regulatory approval is possible. This technology is not yet commercialised but development of commercial plastics with these properties should have a major impact on distribution of a wide range of horticultural produce.

Aldehydes such as hexanal and heptanal are formed when fats in foods such as cereals, dairy products and fish are oxidised. The ongoing pressure to include recycled polyolefins in packaging will increase the need for removal of aldehydes and other oxidation products generated when plastics are given multiple thermal treatments. DuPont (UK) is reported to be supplying multilayer film with 25% post-consumer recycled content (Anon., 1999). Odour and Taste Control Technology™ developed by the same company is based on incorporation of molecular sieves with a pore diameter around 5 nm (Anon., 1996b).

If this approach is proven to remove aldehydes effectively it may be attractive to plastics processors and regulators alike owing to the simplicity of the materials.

The rapid growth in the number of plastics compositions for oxygen scavenging has led to strong interest in absorption of malodorous and other by-products from oxidation of either the polymers or their additive plastics. Frequently an additive that absorbs small by-products has been included in the patent claims. Absorbents claimed include polyimines for acids, zeolites for organics, especially aldehydes, and in some cases just a second laminated functional barrier to odours but not to oxygen.

One of the approaches to oxygen scavenging packaging discussed later in this chapter involves autoxidation of polydiene polymers and this can be expected to result in formation of a variety of small oxidation products. Several of these have been identified and quantified in the case of blends of polyethylene with polybutadiene, polyoctenamer and styrene/butadiene block copolymer (Katsumoto and Ching, 1997) Groups I and II metal oxides, hydroxides, carbonates and bicarbonates, as well as non-migratory amines have been claimed to be suitable neutralising materials for acids and aldehydes. A variety of porous absorbents have also been demonstrated in other cases.

It will be necessary for industry and regulators to ensure that such processes are not used to conceal the marketing of substandard or even dangerous foods and other products if, for instance, microbial odours were to be scavenged.

7.3.3 Addition of desired components

The concept of addition of substances from packaging to foods is not consistent with the current European Union (EU) Directive for food packaging, which prohibits the total release of any substances into foods in amounts greater than 60 mg/kg of food (Ahvenainen and Hurme, 1997). This applies regardless of how beneficial the substance may be to the consumer of the food. Since this Directive was designed to protect consumers from undesirable migrants an EU-sponsored project, ‘Actipack’ (coordinated by TNO, Zeist, Netherlands), has been established to provide evidence for consideration in any amendments that may be made to accommodate active packaging.

Substances that might be released deliberately into packaged goods may be summarised as shown in Table 7.4 which indicates the current status of commercial application in packaging. Although release of antimicrobial agents is the subject of the greatest interest at present there has been very little commercial application outside Japan. Some of the research has been reviewed (Appendini and Hotchkiss, 1997). These compounds have been incorporated into typical food-contact plastics such as polyethylene, ethylene vinyl acetate copolymer or poly(vinylchloride). Agents used include quaternary ammonium compounds, imazalil, various food acidulants and benzoic anhydride. Silver ions have been

Table 7.4 Release of active agents by plastics

Active agent	Example	Polymer	Status
Antimicrobial	Acids	Polyethylene	Commercial
	Fungicide	Polyethylene	Research
	Silver	Polystyrene	Developed
Chemical inhibitor	Sulfur dioxide	Polyolefins	Research
	VCI*	Polyethylene	Commercial
Antioxidant	BHT**	Polyethylene	Research
	Tocopherol	Polyethylene	Commercial
Aromas	Perfumes	Polypropylene	Commercial
Insecticide	Methyl salicylate	Coating	Commercial
	Pyrethrins	Polypropylene	Research

*Volatile corrosion inhibitor
**Butylated hydroxytoluene (2,6-di-tert-butyl-p-cresol)

bound into porous zeolite particles, which in turn have been incorporated into coatings on glass or into polyethylene films.

A range of antimicrobial agents with potential for use in packaging plastics has been listed (Hotchkiss, 1995). Release of permitted food acidulants has been investigated and it has been shown that their polarity makes them unsuitable for use in the common non-polar sealant plastics. However, conversion to anhydrides which are less polar overcomes this problem. Anhydrides are relatively stable to heating involved in plastics processing and can be hydrolysed by water vapour supplied by some foods. The antifungal agent, imazalil, has also been shown to be suitable for release from a low density polyethylene film in quantities that are effective in fruit and vegetables or for cheese.

Non-enzymic browning of products such as dried fruits and wines is prevented by sulfur dioxide in the range from about 15 ppm to 3000 ppm. This prevents development of undesirable colour and flavour caused by the Maillard reaction (Davis *et al.*, 1975; Davis, 1978). The ease of oxidation of many flavours suggests an opportunity to provide slow release flavour precursors in the packaging material. The application of this approach to a wide range of foods has been proposed (Floros *et al.*, 1997). The manufacture of flavour concentrates in the common commodity plastics has been described by Venator (1986). There is therefore an opportunity to replace flavour that might otherwise be lost by permeation, scalping or oxidation during distribution; however, consumer protection regulations may be necessary if the concept were to be used to mask degraded foods and other products judged on the basis of aroma.

Waxed paper has been used as a reservoir for antioxidant release in the US cereal industry (Labuza and Breene, 1989). Such release allows the opportunity to retain a high concentration at the food surface which is exposed to the air. Little interest has been shown in this subject except in edible coating research. However, polyethylene masterbatches containing α-tocopherol are manufactured by Ciba Specialty Chemicals, offering the advantage of a natural food antioxidant in case of migration.

Outward diffusion of an insect repellant has the potential to allow use of commodity plastics in place of more expensive ones which provide a hard surface. The use of methyl salicylate is commercial in RepelKote™ manufactured by Tenneco Packaging (Lake Forest, IL, USA) for use in paperboard boxes (Barlas, 1998). Early work had shown the potential application of a pyrethrin in this role in polypropylene (Rooney, 1995).

7.4 Reactive polymers

7.4.1 Oxygen scavenging

Oxygen scavenging plastics based on reactive polymers are still in their infancy, although the patent literature indicates a widespread commercial interest in their potential by companies such as Chevron Petroleum, Sealed Air Corporation and BP Amoco. Seemingly simple plastics based on blending of low molecular weight reagents with polymers can present regulatory problems but the future seems bright for polymers that are reactive towards oxygen. Since such plastics need to be handled in air, the control of their reactivity by use of an activation step, such as by use of light, appears to offer a convenient way of overcoming this limitation.

Reactive polymers may be required either to scavenge headspace oxygen or to react with it as it permeates the package. The latter effect is described as an enhanced barrier. Close-fitting packages such as vacuum packs for block cheese and meats or aseptic cartons of beverages are examples where the headspace is very small and oxygen permeation is the cause of quality loss. It is in such circumstances that oxygen-scavenging plastics films are particularly needed. The use of the oxygen scavenging reaction to intercept oxygen diffusing through the package wall is an example of a chemical barrier as distinct from the physical barrier normally provided by aluminium foil, vacuum metallising or crystalline polymers such as poly(vinylidene chloride).

However, it is in the field of plastics beverage bottles that the chemical barrier to oxygen permeation will be of most value. Beer flavour is particularly sensitive to oxidation and PET alone does not provide a barrier for more than around a month at ambient temperatures. It is widely accepted that 1 ppm oxygen in a bottle leads to a shelf-life of 3 months or less. This quantity is available in aggregate from that dissolved in the PET, beer and in the headspace after as little as a month's storage. Accordingly, a range of multilayer bottles with middle barrier layers have been produced but these still allow only around 5 months' shelf-life.

It is claimed that some external and internal coatings will overcome the final hurdle to achieve 6–12 months' shelf-life but oxygen scavenging plastics which consume headspace, dissolved and permeating oxygen are in demand to provide a complete solution to the oxygen problem. The dual capabilities of oxygen scavenging polymers as headspace oxygen scavengers and permeation barriers is demonstrated to greatest commercial advantage in this field of packaging.

The introduction of a barrier layer such as MXD-6 nylon, EVOH or the new DOW plastic BLOX into the PET bottle can give 4 or more months' shelf-life, dependent upon the packaging and storage conditions which determine dissolved and headspace oxygen levels. There are commercial claims that external or internal barriers made from vacuum or plasma-deposited internal coatings or cross-linked epoxy polymers will give similar shelf-lives. These barriers reduce oxygen ingress by both permeation and release of some polymer-dissolved oxygen when applied internally. However, they do not address oxygen dissolved in the beer or in the headspace, either initially or by diffusion through the closure liner. It appears that the complete solution to packaging of beer in PET bottles requires a scavenger in the bottle wall and in the closure liner to optimise shelf-life.

Some characteristics of commercial or near-commercial oxygen scavenging plastics based on reactive polymers are summarised in Table 7.5. The chemistry underlying reactive polymers is discussed according to the method of activation. Where water is needed it acts as a reaction solvent or to increase the oxygen permeability of hydrophilic barrier films.

Table 7.5 Development status of polymeric oxygen scavengers

Polymer	Activation	Company	Trade name	Comments
unrestricted	UV light	CSIRO	ZERO2	development
polydienes	metal/UV	CRYOVAC	OS1000	commercial
polydienes-co-PET	metal/UV	BP Amoco	OS3000	commercial
polydienes-co-polyamides	metal/UV	BP Amoco	na	research
polyterpenes	metal/UV	Chevron	na	research
polydienes	metal/UV	Chevron	na	research
ester copolymers	metal	Chevron	na	research
aromatic polyamides	metal	Crown, Cork & Seal	Ox-Bar	suspended
polyamides	?	Continental PET	na	commercial
polyketones	metal/UV	Continental PET	na	research

7.4.1.1 Light-induced reactions

The singlet oxygen reactions of compounds dissolved in plastics described earlier have parallels in the form of reactive polymers. It has been shown that the double bond of natural rubber can be photooxidised at a rate sufficient to bring about rapid oxygen scavenging from the headspace of a package. In one example oxygen was scavenged from air, 20 ml, in an illuminated pouch, 100 cm^2 in area, coated on the inside with either natural rubber dyed with tetraphenylporphine, 10^{-3} M, or ethyl cellulose 0.6 M with respect to bis(furfurylidene)pentaerythritol, a highly reactive furan acetal, and 10^{-3} M with respect to tetraphenylporphine (Rooney, 1982b). The rubber scavenged oxygen substantially faster than did the furan in the ethyl cellulose, probably as a result of the higher concentration of double bonds (13.5 equivalents/l) in the rubber than the concentration of furan rings (1.2 equivalents/l of film) in the furan derivative.

It is significant that the reaction occurring in natural rubber polymer is rapid despite the restricted mobility of the polymerised isoprene. The photooxidation of the tacky natural rubber layer resulted in rapid vulcanisation as indicated by the loss of tack within minutes. The degradation of the rubber continued on dark storage, resulting in formation of a brittle powdery film and the release of a strong rancid odour.

The potential for use of synthetic rubbers in place of natural rubber was investigated in order to determine whether the nature of the rubber monomers affects the rate of scavenging (Rooney, 1995). The rates of oxygen scavenging by *cis*-polybutadiene (PB), *cis*-polyisoprene (PI) and poly(dimethylbutadiene) were compared. Consideration of the reactivity of simple low molecular weight analogues of these polymers, would lead to the hypothesis that the more highly methylated rubber, poly(dimethylbutadiene) (PDMB), would react more rapidly with singlet oxygen. However, the inverse relationship is observed, presumably because of the low oxygen permeability of PDMB of 2.1 Barrers compared with those of 20 Barrers and 24 Barrers for PB and PI, respectively. Thus permeability rather than reactivity can be the rate-limiting variable in an oxygen scavenging polymer system.

Since commercial packages made from oxygen scavenging films need to be stored in darkness during warehousing and distribution, they require some form of activation rather than constant illumination with visible light. The systems which have been reported recently for use in coatings, laminations or other plastics layers are normally activated by one of the following mechanisms:

- UV irradiation for photoreduction of active components
- absorption of water to accelerate oxidation and
- UV irradiation for antioxidant destruction

The first use of UV irradiation involves the photoreduction of quinones and other compounds bound to polymer chains and the subsequent oxidation of the reduced species on exposure to oxygen (Rooney, 1994). This reaction sequence does not require transition metal catalysts but involves the conversion of oxygen into more reactive species, such as hydrogen peroxide, which then reacts with antioxidants in a process designed to minimise damage to the backbone of the polymer.

A different approach has been claimed by Continental PET in their use of polyketones of the type made by Shell Chemical. They utilise the oxidative instability coupled with catalysts, UV irradiation and water absorbed from the packaged product. This process is not unlike the initiation of the photodegradable plastics in the 1970s. The proposed use of such oxygen scavenging polymers involves sandwiching the polyketone between layers of PET, which provides some barrier to oxidation products. Carbon dioxide formed as a result of the polymer oxidation reaction is turned to benefit in carbonated beverage packaging.

7.4.1.2 Metal-catalysed reactions

The first product of this type developed as an oxygen scavenging polymer was named Ox-Bar™ (Crown Cork and Seal Ltd, UK). This process is based on the cobalt-catalysed oxidation of polyamides in which the diamine monomer is aromatic resulting in oxidisable methylene groups which are not present in aliphatic polyamides. The oxidation products are uncertain and this process was not commercialised by its inventors. The specific advantage of this system is the miscibility of aromatic polyamides with PET which, in the case of MXD-6 nylon, allows up to 4% to be blended into the polyester (Folland, 1990). When used with 200 ppm of cobalt as the stearate salt, this polyblend allows blowing of bottles with an oxygen permeability of essentially zero, for one year. In some instances activation might not be a necessary consideration if the package is prepared from all constituents immediately before filling. Such packaging systems would include blow-moulding of beverage bottles which often occurs on the premises of the beverage filler. Thus the catalyst and the oxidisable substrate can be kept apart until the preform is injection moulded as in the Ox-Bar process.

A further factor in the effectiveness of oxygen scavengers is the nature of the packaging material carrying the oxygen scavenger. In the case of the bottles containing MXD-6 nylon blended into the PET, for instance, the oxygen permeability of the bottle is so low that a delay between blowing and closure is quite reasonable. Since much of the beer industry will probably purchase preforms for blowing in-line, the potential for use of such a system seems good.

Several companies have revisited the oxidisable aromatic nylons. Continental PET Technologies Inc. of the US has introduced commercially a 5-layer beer bottle appearing to consist of a central and two outer layers of PET sandwiching two layers of MXD-6 nylon. This package is an effective barrier to oxygen permeation and the nylon layers appear to consume oxygen, giving the shelf-life demanded by the consumer market. The mechanism by which this oxygen barrier is generated needs to be elaborated.

Polyethylene terephthalate (PET) is made from the aromatic terephthalic acid and ethylene glycol. As a result there are methylene groups adjacent to the ethereal oxygen of the ester group, rather than the electron-withdrawing carbonyls. This results in no substantial weakening of the methylenic C—H bonds, compared with those in alkanes, and therefore the bonds are thermally stable. When a polyester is made from a dihydroxybenzene and an aliphatic diacid there are one or more methylenes adjacent to the carbonyls. This is claimed to result in weakening of the C—H bonds to such an extent that the polyester is substantially more oxidisable than PET at room temperature. The results presented by the inventors suggest that use of this type of polymer blended with PET could make a substantial contribution to the packaging of juices and beer since the oxygen barrier is around 77 times higher than that of PET. The authors suggest the use of particular grades of PET which contain cobalt catalyst residues and the cobalt would be expected to play a major role in the autoxidation reactions involved (Schmidt *et al.*, 1997).

7.4.1.3 Polydienes

The oxidation of polydienes can be accelerated by inclusion of soluble transition metal catalysts. It has been shown that poly(1,2-butadiene) containing cobalt octoate can scavenge oxygen from package headspaces by oxidation of double bonds mostly pendant from the polymer backbone (Speer *et al.*, 1993). This process approaches the problem of premature reaction with oxygen by pre-planned activation involving a slow generation of full capacity by consumption of antioxidants. The process is activated by exposure of the polymer to UV light, electron beam or corona discharge energy sources which cause more rapid consumption of the antioxidants and thus reduce the induction period of the free radical autoxidation. This type of film, involving side-chain oxidation of a polydiene or other oxidisable polymers with glass transition temperatures well below 0°C, is designed as both a headspace oxygen scavenger and an oxygen permeation barrier for refrigerated products.

The hydroperoxide formed by reaction of either the polymer free radical or direct attack on the polymer decomposes to form ketones, aldehydes, peroxides or carboxylic acids, in some cases with chain scission. In the case of poly(1,2-butadiene) it is the aim to localise the oxidation, as much as possible, to the side-chain containing the double bond. Speer *et al.* (1993) report that this polymer retains its physical properties even when most or all of its oxidisable groups have been reacted. Since this is a typical autoxidation that is also thermally activated, antioxidants are added to the polymer to stabilise it against premature initiation in the extruder or moulding machine in package fabrication. The quantity of antioxidant is chosen both to provide this protection and to retard the autoxidation chain reaction for a selected period of time between activation and package filling.

This plastic is supplied by the CRYOVAC Company as a co-extrusion with a sealant as a migration barrier and is suitable for lamination to an oxygen barrier film. Proposed applications are for short shelf-life bakery products and processed meats.

The use of such compositions at refrigerator temperatures required for storage of meats and cheese requires that the reactive layer be readily permeable to oxygen. Speer and Roberts (1993) have specified that these compositions should be largely amorphous and have a glass transition temperature below −15°C.

A chemistry similar to that described above has been developed by Amoco Corporation (now BP Amoco) for commercial use in PET plastics such as 3-layer blow-moulded bottles for beer or other beverages (Cahill and Chen, 1997). The copolymer is a block copolymer of PET with a 2–12% polybutadiene or other polyolefin with tertiary or allylic hydrogens and containing from 50 to 300 ppm of cobalt octanoate. The reaction is described as being accelerated by exposure to near-UV light, preferably shortly before the package is filled. The polydiene can be expected to undergo a similar autoxidation reaction sequence to that found with the CRYOVAC composition. However, this copolymer can be expected to be less permeable to oxygen and any small organic oxidation products than in the CRYOVAC case where the polydiene is a simple hydrocarbon polymer. Results so far suggest that the introduction of the polydiene block into the middle layer does not result in substantial damage to the carbon dioxide barrier of a multilayer PET bottle. The company has also applied for a patent for copolyamide compositions with the polyolefin oligomer incorporated into the polyamide chain by reactive extrusion (Cahill *et al.*, 1997). The oxygen scavenging chemistry is similar to that in the copolyester case.

7.4.1.4 Miscellaneous systems

Chevron Chemical Company have developed transition-metal catalysed oxygen scavenging polymers with pendant benzylic esters made from ethylene copolymers with acrylic acid (Ching *et al.*, 1994). The oxidation results in release of benzoic acid which has food additive approval. However, this system requires development of at least a partial functional barrier to this compound if it is to be developed commercially. The inventors have applied for a patent for a variety of by-products of oxidation of polymers used as oxygen scavengers (Ching *et al.*, 1997). In subsequent patent applications the inventors have claimed autoxidation of styrene-butadiene copolymer and, in a separate application, polyterpenes dissolved in polymers also with a transition metal catalyst. The terpenes include food components such as α-pinene and limonene (Katsumoto and Ching, 1997; Katsumoto *et al.*, 1998).

There is a performance-based distinction between packaging structures with the reactive components separated from the product or its headspace by layers of differing oxygen permeability. Those plastics proposed as middle layer in a PET-based multilayer will not be rapid headspace scavengers but will serve largely as a chemical barrier to oxygen ingress. Plastics in which the reactive layer is separated from the product by, at most, a polyolefin heat-seal can be expected to scavenge headspace and dissolved oxygen rapidly, within a few days.

7.4.2 Absorption of odours and food components

The ‘scalping’ of aromas and flavours from foods and perfumed products has long been considered a disadvantage of sealant plastics when compared with barrier plastics and other packaging media (Hirose *et al.*, 1989). The best known of these interactions is sorption of limonene from orange juice. There have been some instances where simple Van der Waals interactions between a plastic and a food component have been shown to be useful in reducing the concentration of the latter in the food (Chandler *et al.*, 1968). This work revealed that limonin, a bitter compound found in orange juice, can be absorbed by cellulose ester polymers. A cellulose acetate-butyrate liner in a bottle removed 75% of the limonin from 500 ml of juice after 3 days at 4°C. This process has considerable potential in the freshly squeezed navel juice market.

Besides the processes for aldehyde removal by means of absorbent particles described earlier in this chapter there is also an approach based on reactive polymers. This research demonstrates removal of aldehydes such as hexanal and heptanal from package headspaces by use of polyimines which contain amine groups pendant from the backbone (Visioli, 1991). The polyimines, preferably polyethyleneimine, are chosen to have minimal branching, thereby maximising the concentration of primary amines capable of forming Schiff bases with aldehydes. This technology was further developed for application in recycled high density polyethylene which contains low molecular weight aldehydes coming from milk residues in the original milk jugs (Visioli and Brodie, 1994). The polyalkylimines are liquids with molecular weights up to 2500 Daltons and are dispersed at a preferred droplet size below 10 μm in a polyethylene master batch. The chance of migration of the reactive polymer is further reduced by incorporation of some acid copolymer which also improves compatibility. This DuPont technology appears not to have been widely promoted since DuPont Odour Control Technology based on porous solids is also available (see elsewhere in this chapter; Section 7.3.2).

7.4.3 Antimicrobial surface generation

The containment of beverages and foods with closely-fitted packaging frequently demands that such packaging is sterilised or has a reduced microbial loading. Most commonly a mist of hydrogen peroxide or other sterilant is used, sometimes in combination with UV irradiation. Since the handling of air containing oxidising agents such as hydrogen peroxide imposes additional engineering requirements on the beverage packer there has been ongoing interest in generation of plastic surfaces with antimicrobial properties.

This field is still a research area involving reactive polymers that are normally designed not to release any of the active agent into the food. Initial work involved surface binding of benomyl, a fungicide, to the surface of Surlyn™ (Dupont), an ionomer film (Halek and Garg, 1989). The fungicide is heat labile and is decomposed in common organic solvents and in some aqueous solutions. It was therefore appropriate to bind it to a polymer in an effort to stabilise it, as well as to test its activity while bound. The common sealant film is made from a copolymer of ethylene with methacrylic acid and one of its metal salts.

The binding process involved soaking the film in a DMF solution of the binding agent for 6 days, followed by reaction with the fungicide for a similar period. This process would need substantial modification if it were to be commercialised. The test results with strains of *Aspergillus flavus* and *Penicillium notatum* were positive inhibition on agar in Petri dish experiments, but appearance of a zone of inhibition beyond the edges of the plastic film indicated some release of benomyl from the polymer matrix. The fungicide is not approved for food use so any future binding would require an improved binding procedure.

A slightly different approach has been taken by Weng *et al.* (1997), who reacted benzoyl chloride with variants on Surlyn 8940 to generate anhydride bonds with the polymer. In this case the antimicrobial surface was generated by release of benzoic acid, an approved food preservative, on hydrolysis of the anhydride bonds by food-derived water. It was found that up to 1.6 mg/g of film could be released, suggesting an effective initial impact on surface contamination may be possible if the process is developed further.

Lysozyme is an enzyme that is particularly effective against Gram-positive bacteria and it has been bound to a variety of solid substrates. Appendini and Hotchkiss (1997) bound this enzyme to nylon 6,6 and poly(vinyl alcohol) and physically immobilised it in cellulose triacetate polymer films. The best results were obtained with the physically immobilised enzyme. The enzyme could be dispersed in the cellulose ester at levels up to 15% by weight and was tested against a suspension of *Micrococcus lysodeiticus* in water. This organism is particularly sensitive to lysozyme and was killed by contact with the film. Reductions in population of 10^7 were observed after 20 hours' exposure using a surface to volume ratio of 1:100. This suggests that application in beverage packaging may be attractive, especially after the detail of the nature of the dispersion in the plastic has been elucidated. However, the work showed that immobilisation of a GRAS food enzyme in a plastic is possible without chemical bonding and that migration was very minor. It would probably be necessary to develop a fabrication method that concentrates the enzyme near the polymer surface to overcome cost as a major impediment to use.

Since the scope for activity of enzymes is normally limited, it will probably be necessary for substantial research to expand this pioneering work in order to determine whether the range of microorganisms killed meets industry standards.

Besides the incorporation of known antimicrobial agents into or on to plastic surfaces there has been ongoing research into chemical modification of surfaces to generate chemical groups toxic to microorganisms. Some work has indicated that acid groups present on the natural polymer chistosan, separated from crustacean shell waste, could be used in this way (Chen *et al.*, 1996). However, the approach of rapid conversion of chemical groups on the surface of extruded plastics packaging is very attractive from an industrial viewpoint. In this way the conversion of polymers to their active form may only need to be achieved in a surface layer.

Surface nitrogen-containing groups on the surface of a variety of polymers have been converted to amines by subjecting preformed films to strong light from a laser at 193 nm. Exposure of films treated in this way to *Staphlococcus aureus*, *Pseudomonas fluorescens* and *Enterococcus faecalis* suspensions in phosphate buffer resulted in a decrease in microbial count by over 4, 2 and 1 logs, respectively (Paik *et al.*, 1998).

There have been some discussions as to whether surface-bound amine groups are capable of causing cell death. However, results from reinoculation experiments suggest that cells adsorbed are either dead or unable to reproduce after adsorption. If the microbial kill results can be corroborated, the approach appears attractive for commercial use. Application to food packaging would require investigation of the possibility that migrating species are formed by the irradiation. Photoprocessing is a practical way in which to treat preformed plastics films as is demonstrated by corona-discharge treatment routinely used in activating commercial films before printing or lamination.

7.4.4 Physical property modification

7.4.4.1 Permeability modification

The ratio of the permeability of plastics to a variety of gases and vapours can be a critical factor in the packaging of horticultural produce. In the simplest case of oxygen and water vapour the produce requires a restricted oxygen supply for respiration control, while the water vapour transpired by the produce should not be allowed to condense on the package on cooling. The latter problem could be solved if the water vapour permeability of the plastic allowed water to escape without allowing an oversupply of oxygen. Unmodified plastics do not normally satisfy these two requirements and some form of active packaging is needed to achieve optimum product condition.

The importance of predicting the relative permeability requirements of plastic films to carbon dioxide and oxygen in packaging of fresh horticultural produce has been discussed by Yam and Lee (1995). Various semipermeable patches and techniques for generating controlled porosity have been developed without providing the atmospheric control required, especially in the distribution of fresh-cut salads for retail sale.

This field of generation of selectively permeable plastics remains the most challenging problem for horticultural packaging. Besides the absolute value of the permeability and the various permeability ratios, there is a need for the permeabilities to oxygen and carbon dioxide to respond to temperature change in the same manner as does the produce's respiration rate.

The approach taken by Landec Corporation (Menlo Park, CA, USA) offers promise if their Intellimer™ polymers can be manufactured more economically and with a wide range of gas permeability ratios, in addition to the temperature abuse properties which they already demonstrate. These copolymers are based largely on ethylene with acrylic acid esters of long chain aliphatic acids, such as fatty acids. The side-chains crystallise and the melting point can be varied with a precision of $\pm 2^\circ$C (Stewart *et al.*, 1994). At the side-chain melting point the gas permeability increases sharply and the response to further rise in temperature can also increase. The value of this response is to make these films suitable for rapidly respiring produce such as cut cabbage. The polymers can be used as pouches or as patches to be placed over holes in packages made from commodity polymers.

7.4.4.2 Microwave absorption and reflection

The vacuum metallising of PET film to a level of around 50% optical transmission results in the formation of microwave susceptor properties (Sacharow, 1995). The aluminium layer is discontinuous at this loading, resulting in an array of resistors which heats when exposed to microwave energy. The lamination of such films to paperboard led to useful packaging materials for crisping of pastries, frozen meat pies and, most commonly, for popping corn. The background to this field has been discussed (Sacharow and Schiffman, 1992).

Besides merely heating a packaged product, microwaves can selectively heat mixed products, such as TV dinners, in which some components may need to be kept unheated while others are required warm. Shielding by continuous layers and field intensification by using patches of metallising have been developed but little used. Early work by the Alcan company resulted in a product termed MicroMatch™, which had selectively metallised domes designed to cover the food to be heated. The full potential of microwave susceptor packaging will probably not be reached until more acceptable replacements for aluminium are developed. Aluminium oxidises when heated in air, resulting in loss of susceptor performance.

7.5 Plastic composite structures

7.5.1 Absorption of water

Packaging films are often chosen for their high barrier to water vapour. There are, however, some packaging situations in which successful storage of foods requires water absorption or some control of relative humidity. The circumstances in which water may need to be controlled in food packs include:

- *accumulation of liquid water*, as in
  melting of ice used in fish transportation,
  tissue fluids from cut fresh meats, fruits and vegetables, and

- *condensation of water vapour*
  from high water activity (a_W; % relative humidity expressed on a scale 0–1) foods undergoing temperature cycling, and transpiration of living fruit and vegetables.

Packages of these products can have a build-up of liquid water which endangers microbiological stability or at least looks unattractive. Moisture migration via the vapour phase can result in transient formation of regions without adequate preservative or where the a_W is high even though the food is packed at lower a_W. This water can be absorbed by packaging structures acting as humidity buffers.

Temperature cycling of packages of high water activity foods has led most film manufacturers to offer films for fruit and vegetable packaging with an anti-fog additive. These amphoteric additives are blended with the resin before extrusion and migrate to the surface after film formation. The result is a lowering of the interfacial tension between water condensate and the plastic film causing the droplets of water to coalesce. The film of water may drain and be absorbed as above or aggregates at the lowest point in the package. Anti-fog treatments are a cosmetic form of active packaging, assisting the customer to see the packaged food clearly.

7.5.1.1 Humidity buffering to absorb water vapour

Water vapour can be intercepted by a plastic sorbent layer provided the latter has a sufficiently high capacity and water vapour diffusion coefficient. A recent development for distribution of produce involves a water-barrier coating of the inside of fibreboard cartons to allow moist produce to be placed directly into the carton, and a water-vapour-sorbent film layer which absorbs water at high relative humidity, such as when the temperature decreases. When the temperature rises the water can evaporate from the sorbent and be reabsorbed by the produce. Such packages are described as humidity buffers. Since temperature cycling is very difficult to avoid during handling, there is normally every likelihood of condensation in produce packs and with this the growth of microorganisms on fruits and vegetables. The need to avoid condensation is one of the main impediments to the introduction of equilibrium modified atmosphere packaging in sealed packs.

The humidity buffering designs of Patterson and Joyce (1993) involve: a) an integral water vapour barrier layer on the inner surface of the fibreboard; b) a paper-like polymer material bonded to the barrier which acts as a wick; and c) a layer highly permeable to water vapour but unwettable next to the fruit or vegetable. The latter layer is spot welded to the layer underneath.

This system has been demonstrated to prevent condensation of water in cartons containing carrots, 6 kg, which were cooled to 3°C from 10°C. After 3 days the amount of free water in the humidity buffered carton was negligible, whereas the reference carton contained 10 g water. The results demonstrate the capacity of such a system for interception of water vapour under practical conditions.

A somewhat related system for use without a paperboard carton is marketed under the name Pitchit™ and manufactured by Showa Denko in Japan. The structure is a duplex of two layers of poly(vinyl alcohol) between which is a liquid layer consisting of an alcohol, described by Labuza and Breene (1989) as propylene glycol, and a carbohydrate, both of which are humectants. The speed of water uptake of this structure has been reported (Rooney, 1995).

Pitchit is marketed for home use in roll or single sheet form for wrapping pieces of meat or fish to reduce the a_W of the atmosphere surrounding the food. Louis and de Leiris (1991) report the availability of an additional moisture control film which plays the same role as Pitchit in the Japanese market.

It has been reported that polyvinyl alcohol blankets, apparently without any additional constituents, have been manufactured in the US by W.R. Grace Ltd (now Sealed Air Corporation) (Floros *et al.*, 1997).

7.5.1.2 Liquid water control

It is now becoming common for flesh foods which weep juice on storage to be placed on drip-absorbent sheets. Several companies manufacture these sheets which basically consist of two layers of a microporous or non-woven film, like polyethylene or polypropylene, between which is placed a superabsorbent polymer powder. The duplex sheet is sealed at the edges and in a cross-hatch pattern to prevent conversion of the sheet into a large ‘pillow’ when considerable amounts of water have been absorbed. Large sheets are used for absorption of melted ice in the packaging of seafood for air transportation. The superabsorbent polyacrylate salts used in such sheets are capable of absorbing between 100 and 500 times their own weight of liquid water.

Absorbent sheets have been used in a more complex structure in the Verifrais™ process of the Codimer company. In this process pads containing organic acid and sodium bicarbonate are placed in the lower compartment of a polystyrene meat tray, obscured from view. The two compartments are joined by holes in the polystyrene. Juice from the meat passes through the holes, dissolving the two reagents which form carbon dioxide. This helps to inhibit microbial growth on the meat. The rate of gas formation is controlled to some extent by sheets of absorbent paper that delay the passage of the water to the reagents (Louis and de Leiris, 1991).

7.5.1.3 Oxygen absorbers in plastics sachets

The most common oxygen absorbers used in the food and pharmaceutical industries consist of porous sachets containing iron powder mixed with other ingredients to assist the process of rusting. These sachets frequently consist either of non-woven polyethylene or of other plastics which have been made porous after film extrusion. Polyester is frequently used in this role. It is important that the sachets are resistant to tearing as spillage of the iron powder would result in rejection of the contents of the package in which the sachet is placed.

The problem of spillage of the iron powder is being addressed by Mitsubishi Gas Chemical Company by incorporating the iron into strips of commodity plastics which are then placed in the familiar sachets. These sachets are also supplied in the form of adhesive 'labels' or heat-sealed to closure liners for jars. These approaches help to overcome the perception that the sachet has been added to a food. The subject of sachets and their use in packaging has been reviewed (Smith *et al.*, 1995).

7.5.1.4 Thermal protection

Active packaging plastics systems are used or proposed for use in packages which self-heat, self-cool or provide thermal insulation by physical or chemical means. There have been several approaches to providing thermal buffers by means of double-walled container construction involving gels. However, a recent proposal for use in cups for cool take-away beverages utilises molecular-alloys with high latent heat of phase transition (Espeau *et al.*, 1997). This is analogous to the use of the ice-water phase transition but the chosen alloys of hydrocarbons melt at temperatures from 6 to 13°C. Optimal shapes of drinking and convenience vessels were investigated and it was found that an 'isothermal water bottle' could keep a drink at temperatures between these limits for over 3 hours.

7.6 Conclusions

Active packaging plastics for use in packaging applications have been elevated from a scientific curiosity to a small, but rapidly expanding, field of development and manufacture in two decades. The growth has been uneven, with compositions for oxygen scavenging in food packs being the main focus. The uptake of the concepts will depend strongly upon whether the additional cost often associated with active materials is justified in comparison with passive alternatives. At this stage of development the regulatory and environmental hurdles have not been well discussed, especially where chemical reactions are involved in achieving activity. The early success of some systems of this type leads to optimism that the field will continue to grow into a mainstream aspect of packaging technology.

References

Ahvenainen, R. and Hurme, E. (1997) Active and smart packaging for meeting consumer demands for quality and safety. *Food Additives and Contaminants*, **14** 753-763.

Anon. (1996a) Oxygen-absorbing packaging materials near market debuts. *Packaging Strategies*, Packaging Strategies Inc., West Chester, PA, USA 1-3.

Anon. (1996b) Odour Eater. *Packaging News*, August, 3.

Anon. (1999) DuPont offers 25% PCR film. *Packaging Strategies*, **17**, December 15, 6.

Appendini, P. and Hotchkiss, J.H. (1997) Immobilization of lysozyme on food contact polymers as potential antimicrobial films. *Packag. Technol. Sci.*, **10** 271-279.

Barlas, S. (1998) Packers tell insects; Stop bugging us! *Packaging World*, **5** (6) 31.

Cahill, P.J. and Chen, S.Y. (1997) *Oxygen scavenging condensation polymers for bottles and packaging articles.* PCT Application PCT/US97/16712.

Cahill, P.J., Richardson, J.A. and Wass, R.V. (1997) *Copolyamide active-passive oxygen barrier resins.* PCT Application PCT/US98/02991.

Chandler, B.V., Kefford, J.F. and Ziemelis, G. (1968) Removal of limonin from bitter orange juice. *J. Sci. Food Agric.*, **19** 83-6.

Chen, M.-C, Yeh, G.H.-C. and Chiang, B.-H. (1996) Antimicrobial and physicochemical properties of methylcellulose and chitosan films containing preservative. *J. Food Proc. Preservat.*, **20** 379-390.

Ching, T.Y., Katsumoto, K., Current, S. and Theard, L. (1994) *Ethylenic oxygen scavenging compositions and process for making same by esterification or transesterification in a reactive extruder.* PCT Application PCT/US94/07854.

Ching, T.Y., Goodrich, J.L. and Katsumoto, K. (1997) *Oxygen scavenging system including a by-product neutralising material.* PCT Application PCT/US97/03307.

Davis, E.G. (1978) The performance of liners for retail wine casks. *J. Food Technol.*, **13** 235-241.

Davis, E.G., McBean, D. McG. and Rooney, M.L. (1975) Packaging of foods that contain sulfur dioxide. *CSIRO Food Res. Q.*, **35** 57-62.

Espeau, P., Mondieig, D., Haget, Y. and Cuevas-Diarte, M.A. (1997) 'Active' package for thermal protection of food products. *Packaging Technology and Science*, **10** 253-260.

Floros, J.D., Dock, L.L. and Han, J.H. (1997) Active packaging technologies and applications. *Food Cosmetics and Drug Packaging*, January, 10-17.

Folland, R. (1990) Ox-bar. A total oxygen barrier system for PET packaging. *Proceedings Pack Alimentaire '90.* Innovative Expositions, Inc., Princeton, NJ, Session B-2.

Halek, G.W. and Garg, A. (1989) Fungal inhibition by a fungicide coupled to an ionomeric film. *J. Food Safety*, **9** 215-222.

Hirose, K., Harte, B.R., Gaicin, J.R., Miltz, J. and Stine, C. (1989) Sorption of d-limonene by sealant films and effects on mechanical properties. In: *Food and Packaging Interactions* (ed. J.H. Hotchkiss), ACS Symposium Series No. 365, American Chemical Society, Washington, DC, pp. 28-41.

Holland, R.V. (1992) *Absorbent materials and uses thereof.* Australian Patent Application No. PJ6333.

Hotchkiss, J.H. (ed.) (1989) *Food and Packaging Interactions*, American Chemical Society, Washington DC.

Hotchkiss, J.H. (1995) Safety considerations in active packaging. In: *Active Food Packaging*, (ed. M.L. Rooney), Blackie A & P, an imprint of Chapman and Hall, Glasgow and London, pp. 238-255.

Katsumoto, K. and Ching, T.Y. (1997) *Multi-component oxygen scavenging composition.* PCT Application PCT/US97/13015.

Katsumoto, K., Ching, T.Y., Goodrich, J.L. and Speer, D. (1998) *Photoinitiators and oxygen scavenging compositions.* PCT Application PCT/US98/07734.

Koros, W.J. (1990) Barrier Polymers and Structures: Overview. In: *Barrier Polymers and Structures* (ed. W.J. Koros), American Chemical Society, Texas, pp. 1-21.

Labuza, T.P. and Breene, W.M. (1989) Applications of 'Active Packaging' for improvement of shelf-life and nutritional quality of fresh and extended shelf-life foods. *J. Food Proc. Preservat.*, **13** 1-69.

Louis, P.J. and de Leiris, J.P. (1991) *Active Packaging*, International Packaging Club, Paris.

Maloba, F.W., Rooney, M.L., Wormell, P. and Nguyen, M. (1996) Improved oxidative stability of sunflower oil in the presence of an oxygen-scavenging film. *J. Amer. Oil Chem. Soc.*, **73** 18-185.

Paik, J.S., Dhanasekharan, M. and Kelly, M.J. (1998) Antimicrobial activity of UV-irradiated nylon film for packaging applications. *Packag. Technol. Sci.*, **11** 179-187.

Patterson, B. and Joyce, D.C. (1993) *A package allowing cooling and preservation of horticultural produce without condensation or desiccants.* PCT Application PCT/AU9300398.

Rooney, M.L. (1982a) Oxygen scavenging from air headspaces by singlet oxygen reactions in polymer media. *J. Food Sci.*, **47** 291-4, 298.

Rooney, M.L. (1982b) Oxygen scavenging: a novel use of rubber photooxidation. *Chem. Ind.*, 197-198.

Rooney, M.L. (1994) *Oxygen-scavenging plastics activated for fresh and processed foods.* Abstracts Annual Convention of the Institute of Food Scientists and Technologists, Atlanta, Georgia.

Rooney, M.L. (ed.) (1995) *Active Food Packaging*, Blackie A & P, an imprint of Chapman & Hall, Glasgow and London.

Rooney, M.L. and Holland, R.V. (1979) Singlet oxygen: an intermediate in the inhibition of oxygen permeation through polymer films. *Chem. Ind.*, 900-901.

Rooney, M.L., Holland, R.V. and Shorter, A.J. (1981) Removal of headspace oxygen by a singlet oxygen reaction in a polymer film. *Journal of the Science of Food and Agriculture*, **32** 265-72.

Sacharow, S. (1995) Commercial applications in North America. In: *Active Food Packaging* (ed. M.L. Rooney), Blackie A & P, an imprint of Chapman & Hall, Glasgow and London, pp. 203-214.

Sacharow, S. and Schiffman, R.F. (1992) *Microwave Packaging*, PIRA International, Leatherhead, 119-134.

Schmidt, S.L., Agrawal, A.S. and Coleman, E.A. (1997) *Transparent oxygen-scavenging article including biaxially-oriented polyester.* PCT Application PCT/ US97/16826.

Smith, J.P., Hoshino, J. and Abe, Y. (1995) Interactive packaging involving sachet technology. In: *Active Food Packaging* (ed. M.L. Rooney), Blackie A & P, an imprint of Chapman & Hall, Glasgow and London, pp. 143-172.

Soares, N.F.F. and Hotchkiss, J.H. (1998) Naringinase immobilization in packaging films for reducing naringin concentration in grapefruit juice. *J. Food Sci.*, **63** 61-65.

Speer, D.V. and Roberts, W.P. (1993) *Improved oxygen scavenging.* PCT Application, US 93/09125.

Speer, D.V., Roberts, W.P. and Morgan, C.R. (1993) *Methods and compositions for oxygen scavenging.* US Patent 5211875.

Stewart, R.F., Mohr, J.M., Budd, E.A., Lok, X.P. and Anul, J. (1994) Temperature compensating films for modified atmosphere packaging of fresh produce. In: *Polymeric Delivery Systems—Properties and Applications* (eds. M.A. El-Nokaly, D.M. Pratt and B.A. Charpentier), ACS Symposium Series No. 520, American Chemical Society, Washington DC.

Teumac, F.N. (1995) The history of oxygen scavenger bottle closures. In: *Active Food Packaging* (ed. M.L. Rooney), Blackie A & P, an imprint of Chapman & Hall, Glasgow and London, pp. 193-202.

Thomas, J.A. (1998) A polymeric oxygen scavenging system and its packaging application. Paper presented at *1998 Annual Meeting and Food Expo.*, Institute of Food Technologists, Atlanta, Georgia, June 24, 11 pp.

Tsai, B. (1996) Amosorb: oxygen scavenging concentrates for package structures. Proc. Future-Pack '96, 11 pp.

Turro, N., Chow, M.F. and Blaustein, M.A. (1981) Generation, diffusivity and quenching of singlet oxygen in polymer matrices investigated by chemiluminescence methods. *J. Chem. Phys.*, **85** 3014.

Venator, T.E. (1986) Fragrance and flavour interactions with plastics packaging materials. *Packaging*, October, 30.

Venkateshwaren, L.N., Chokshi, D., Chiang, W.L. and Tsai, B.C. (1996) *Oxygen scavenging composition.* PCT Application PCT/US96/06721.

Visioli, D.L. (1991) *Novel packaging compositions that extend the shelf-life of oil-containing foods.* US Patent Application Serial No. 07/724,421.

Visioli, D.L. and Brodie, V. (1994) *Method for reducing odors in recycled plastics and compositions relating thereto.* US Patent 5,350,788.

Weng, Y.-M., Chen, M.-J. and Chen, W. (1997) Benzoyl chloride modified ionomer film as antimicrobial food packaging material. *Internat. J. Food Sci. Technol.*, **32** 229-234.

Yam, K.L. and Lee, D.S. (1995) Design of modified atmosphere packaging for fresh produce. In: *Active Food Packaging* (ed. M. L. Rooney), Blackie A & P, an imprint of Chapman & Hall, Glasgow and London, pp. 55-72.

Zagory, D. (1995) Ethylene-removing packaging. In: *Active Food Packaging* (ed. M.L. Rooney), Blackie A & P, an imprint of Chapman & Hall, Glasgow and London, pp. 38-54.

8 The changing image of plastics packaging in the marketplace

A. Streeter

8.1 Introduction

The use of plastics in consumer packaging has enjoyed a rapid rise since the 1950s. Conversion processes such as moulding and co-extrusion have evolved and new technologies have been introduced. This process, coupled with a better understanding of the commercial applications of plastics packaging, has accelerated throughout the 1990s to a point where, at the start of the new millennium, plastics are the dominant force and focus for change in consumer packaging.

This chapter examines some of the reasons why plastics is a dominant force, in the context of global and regional applications.

8.2 The switch to plastics packaging

Consumer packaging is a complex subject involving many disciplines. Above all else, however, it is driven by a consideration of costs. In this area of packaging there are relatively few revolutions, but there is a vigorous culture of evolution often driven by issues of cost. Some commercial forces are considered below.

Unit cost reduction could be achieved by:

- lowering the weight,
- increasing throughput speeds,
- a device to de-skill and reduce direct-labour content.

Enhancing the brand values could be achieved by:

- better decoration,
- differentiation through shaping, and
- adding value through on-pack service features (e.g. easy-open devices).

In both examples, the cost *vs* value scenario has encouraged a stronger end result (value) at a lower cost base. Note that this is not the same as a lower price base; that can only come about by the power of competition.

We now consider the manufacture of other packaging materials. The basic group of materials and their manufacturing requirements are examined below:

Aluminium sheet and cans require:

- for aluminium production, large numbers of electrically powered smelters that are often built close to hydroelectric stations or power generators;
- good access for the raw material, bauxite, to be transported and the end-product to be distributed;
- heavy plant for the manufacture of aluminium sheet;
- heavy presses for pressing and decorating the containers, typically high-speed machines;
- lighter equipment, for rolling aluminium sheet into a foil.

Steel requires:

- heavy equipment for pressing;
- some of the heaviest printing machines ever built for decorating.

Glass containers require:

- a lot of energy for continuous running;
- good transport access for the raw materials silica (sand), and maybe cullet, and for distribution of finished bottles and jars.

Board requires:

- a mill close to sources of raw material, normally a forest of softwood cultivated for this purpose, if manufacturing virgin board, or good transport for waste materials, if mixed materials board;
- large quantities of energy and extensive water reserves;
- a 3-phase electricity supply and some lighter machinery (in relation to that above) for the conversion of board, into cartons, for example.

By comparison, converting the plastics resin is often relatively simple. Certainly the cracking of the oil and the conversion of the plastics resin is a strategic (often national) investment but once made as chips, granules or powder, the resin is clean, stable and free-flowing. Conversion into packaging involves processing by machines that, once set, will continue to perform to order, manufacturing the item or sheet of material with low (human) skill procedures.

Given electricity, water and a building, some quite sophisticated plastics packaging can be produced anywhere in the world. Many a jungle clearing or harsh inhospitable place has an injection moulding or film blowing plant in a simple factory, making packaging for local manufacturers.

This means that contemporary standards of plastics packaging are often readily available locally in many parts of the world, and this leading form of packaging is not simply the preserve of multinational product or brand manufacturers, or densely populated first world societies. This is an important feature of plastics packaging proliferation because, as demonstrated earlier, other packaging materials require heavy plant and prodigious quantities of energy or materials.

The switch to and, more importantly, the acceptance of plastics packaging in many, if not most, of the world markets arises from more than just cost and simplicity of operation. It is the result of many factors converging.

8.2.1 Cost

Packaging cost is a strategic issue. It should be remembered that, in simple terms, plastics is a low density material. Lighter than water, it can be formed by a number of methods, often very 'thin-walled' indeed, and still be useful. Film, for example, as a 30-μm web of orientated polypropylene, can be formed into a pouch, fully sealed and can contain and protect the product inside, all at a very low cost.

That cost is not simply unit cost of the (in this case) film sheet material employed, it is also a combination of key factors. These include:

- low unit weight
- quite high-speed printing of the film
- the ability to be efficiently transported in bulk to point of use (in this case as reels)
- quite high speed transformation from a continuous web into a sealed unit on a form-fill-seal basis
- a barrier to prevent product deterioration and
- a finished pack adequate for distribution, retail display and for the consumer to use

This combination of important and principally critical issues makes plastics packaging so important in commerce. This applies to other plastics packaging:

- simple blow-mould bottles offer style, containment, protection, lightweight and low cost
- injection moulded closures can be complex in shape, precise, protective, lightweight and low cost
- complex bottle moulding such as stretch blow-mould to form PET bottles offers containment, beneficial features like being a pressure vessel, quality exterior finish, and again a low unit cost and
- pressure formed (sometimes know as vacuum formed) delivers rigid containers that are low cost and effective to use in selected markets (e.g. tubs for salads)

Except for injection moulding, these examples can also be undertaken in-line by the product/brand manufacturers; this is particularly true in the soft drinks industry for PET stretch-blown bottles, thereby reducing cost even further. There are a few product manufacturers who injection mould packaging components as part of an in-line packaging operation. This is limited and represents a substantial commitment by the product manufacturer. The cost benefits of plastics packaging force commerce to use it for consumer packaging.

8.2.2 *The environmental agenda*

In nearly every circumstance, a lot of the reduction in costs equates to a reduction in material, which can justifiably equate to a benefit for the environment. To simplify this, the equation/statement is:

> Cost reduction and material reduction go hand in hand; thereby the environment is the nominal benefactor.

Consequently, if we consider plastics and other packaging over the past 30 years, commerce has significantly reduced the quantities of packaging used for any given product. At the same time the number of products available, and bought, has grown considerably, and social standards may have slipped, making packaging waste prevalent and overt; therefore:

> (Product diversity) × (Numbers bought) × (Social attitudes) = More litter. . .
> . . .even if the quantity (weight etc.) of packaging per item has reduced significantly.

In this growth of product availability, plastics has played a significant role; its cost is low and has facilitated many of today's products, but the material is virtually indestructible.

Environmental problems with plastics packaging arise because it is not biodegradable. Although plastics packaging has significantly improved quality of life at minimal cost, society's attitude and lack of infrastructure to control the complete handling on a cradle-to-grave basis has created a strong negative attitude to this medium. Each society or country responds to this differently. An analysis of a selected group of countries on a care/not care basis might look something like the schematic plan shown in Figure 8.1.

This is a generalisation but it does help us to understand the various attitudes towards plastics packaging in different cultures. We shall see further in this chapter how different countries react and use plastics packaging as a consequence of these forces.

8.2.3 *Barrier*

Once packaging can contain, the next critical need is the provision of barrier properties and it is this aspect that represents so much development of

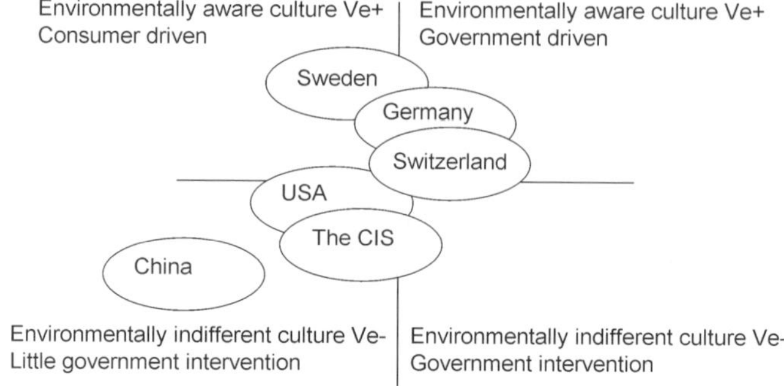

Figure 8.1 A selected group of countries on an environmental care/not care basis.

the science in plastics packaging. There are only three barrier materials in packaging:

- glass, as bottles and jars;
- metal, as steel with coatings or aluminium, both as rigid containers, with aluminium also having an important role as a thin sheet supported by another medium (invariably plastics);
- plastics in various forms.

There are also combinations of these three media. Glass and metal are nominally absolute in their protective qualities whereas plastics developers can increase barrier towards lower and lower transmission rates but never quite reach zero.

8.2.4 Shape

Shaping is important as a brand communication device and as a means of differentiation. Shape can also be protected by design registration and patents and this makes it a particularly valuable feature.

Some plastics packaging can be the most easily shaped packaging in the world, while also providing an inherent, adequate barrier. Its great competitor, in some sectors, is glass but this is relatively limited as costs, market image and other factors apply.

All conversion methods of plastics packaging can provide shape. Even plastic pouches, as we will see, have some shaping possibilities. It is therefore an exceptionally powerful packaging medium.

This ability to shape more freely than glass or metal while delivering barrier (all within reason and technical support) is highly prized by marketing personnel; it has enabled brand growth and enhanced communication.

The application of shaping in plastics packaging is related to the evolved state of the marketplace in which the pack has to work. In underdeveloped societies, packs are apt to be simple, as is the local supply of the packaging; in more advanced societies there is a need for much greater differentiation and the pack is more sophisticated.

8.2.5 Decoration

As markets have become more sophisticated, so the demand for improved decoration has come into being and this can be made very powerful by coordination with pack shape. Irrespective of the printing technology, print standards are always higher off-line in sheet form than on-line or around a shape, such as an aluminium beer can.

Plastics packaging has several, commercially different advantages here too. Flexible packaging is inherently a web material that is also relatively stable, and so it lends itself ideally to rotary printing, at speed and to high standards. Flexible packaging can benefit totally from the printer's art; there is no need to compromise, subject of course to cost, but even here it is often about origination investment not about direct, simple unit cost; a 10 station gravure press is expensive on cylinders but once there the results can be truly magnificent. Lower cost flexographic printing systems, alternatives and advancing technologies are available; excellent results can be had from 4-colour flexo press.

Flexible plastic's effectiveness as a print substrate is exploited commercially in several ways, especially as pouches, giving them authority and value. This has allowed brands to switch format; Figure 8.2 shows Nescafé brands in Europe in freestanding pouches, all decorated to a high standard. Selection of the plastics substrate gives a basis for texture and strength (in consumer perception terms). A credible brand can be marketed in a pouch by manipulating substrate finish—glossy, opalescent, opaque, see-through, metallised, and occasionally a mixture of some of these—with printing inks that can be glossy, matt, opaque or transparent. There are in addition and increasingly, special effects like holographic decoration. The shaping of pouches is discussed later but at this stage the focus is on the commercial power of printing on web-based plastics.

Figure 8.2 Nescafé brands in Europe in freestanding pouches.

A similar argument can be made for horizontal form-fill-seal (commonly known as flow-wraps). Globally impressive executions of this format have promoted brands and products; the confectionery market is a good example.

Both formats benefit from the excellence of finished pack manufacturing, partly resulting from machine development in material control, forming and sealing, and partly from enhanced sealing mediums like cold seal and the increase in web direction stiffness.

The technology of placing a printing sleeve or tube over a rigid container has revolutionised many consumer goods sectors, especially the market drinks. Glass bottles have developed strong personalities; plastic containers have lost their plastics image, hijacked by the dynamics of the final, sleeved container presentation designers exploit the features that a plastics sleeve (and shaped container) have to offer.

There are three significant applications of flexible packaging that have made huge commercial gains from effective print decoration.

The decoration of rigid plastics containers is not so effective in absolute terms, although the technologies are impressive and developing. Injection moulding remains an option with hot foil stamping and tampo printing. Surface coverage is not all that good, but the tooling lends itself to embossing, de-bossing and other surface treatments. The application of these treatments has grown in consumer goods packaging, principally to enhance usage features and add aesthetic values.

The one major exception to this is in-mould labelling, the application of a printed, plastics label in the tooling to form a homogenised weld to the body of the container. The result offers a superb standard of surface decoration with the added benefit (most of the time) of being a single substrate material, allowing recycling as a single substrate but with excellent decorative standards.

Rotary printing of a container is still much used especially in plastic pots, epitomised by yoghurt containers, and globally but to wildly varying standards, influences the brand values attainable. There are still millions of stickers applied to plastics packaging every day because direct print is not commercially available (and also to cope with print errors and promotional needs).

The application and perception of plastics packaging varies by society and industry standards and resources; this is particularly true for surface decoration. We must not forget that design through the medium of print, whatever the method, is the prime communicator and as such has a special role in plastics packaging.

8.2.6 *SDO (simplicity, diversity and opportunity)*

The reasons for the switch to plastics packaging and the effect this has on image and trends in the marketplace could be summarised by stating that plastics packaging offers a lot of SDOs: simplicity, diversity and opportunity.

Simplicity:

- many manufacturing processes and with a low skill procedure,
- well proven technology supported by quality equipment from many sources,
- clean and relatively stable, and
- barrier provider.

Diversity:

- in a variety of conversion processes leading to. . .
  . . .a substantial range of packaging formats,
- can be tailor-made for specific applications, e.g. self-colour or as a closure, and
- has technically strong environmental credentials, if not so strong in society's general perception.

Opportunity:

- a big demand by product manufacturers,
- adds value on a large scale,
- extends shelf-life throughout the supply chain, and
- can be produced in volume.

These factors interrelate and can be considered linked as demonstrated in Figure 8.3.

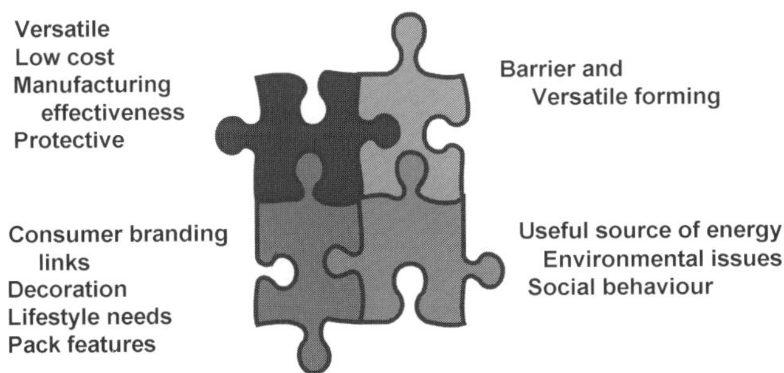

Figure 8.3 The switch to plastics packaging jigsaw-matrix.

8.3 The evolution of lifestyle products

Where does the product end and the packaging start? The evolution of packaging in many of the world markets has reached a level where it is an integral part of the product (a means to add value, offer consumer convenience and so on) but there is a stage beyond that where the pack becomes a badge, a statement about the person or a symbol of beliefs and values.

Packaging as a lifestyle product statement has been particularly effective in drinks and has forced the growth of the market for carbonated soft drinks. Coca-Cola, naturally a powerful player, reintroduced its shaped bottle as a protectable (three-dimensional trademark) shape to reacquire its values for Coke, Coca-Cola brand, in a market of increasing mimics facilitated by common packaging formats, in this case the aluminium can. But today the Coca-Cola bottle (and its now different competitors) is a status symbol, a fashion product, highly visible, portable, easy to use, but more than that, it has emotional value.

This is not a new concept. What is fragrance or perfume but a very high percentage of alcohol and small quantities of a complex mix of aromas in a bottle of extravagant design and cost, way beyond the technical parameters of protection and containment? But then a perfume is all about image, carefully crafted to fulfil an array of interwoven emotions.

That basic premise of lifestyle and emotional values in consumer packaging (and which has forever been an essential ingredient of the perfumes market) has been extended into a whole spectrum of market sectors. This moves or evolves the role of consumer packaging upward, making its lifestyle capabilities critical; consequently the traditional purpose of packaging to protect, contain, provide legal/moral/graphic communication, has become commercially secondary but remains, of course, a technical necessity.

The pyramid in Figure 8.4 simplifies a more complex situation, but it is helpful in understanding where plastics packaging fits in any one region, market or sector; a working method that can be further subdivided into the different plastics packaging options.

It must be remembered that the various roles or society's perceptions are a complex mix of:

- the evolutionary stage of plastics packaging manufacture
- the packaging format and its manufacturing method
- the infrastructure of the region

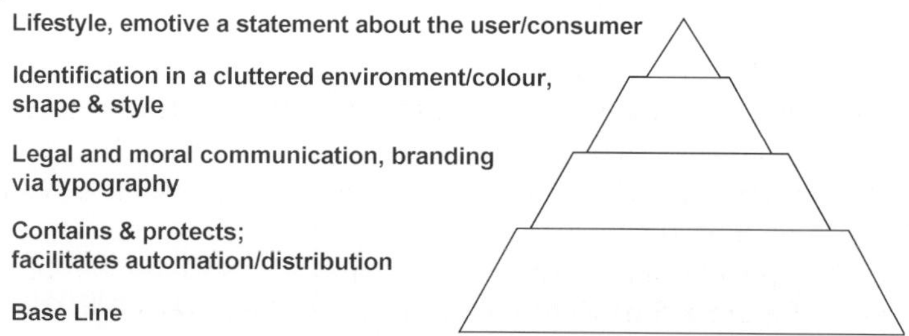

Figure 8.4 The suggested evolution of the role of plastics packaging.

- the evolutionary stage of that society and
- the nature of the market sectors in which plastics packaging operates

In considering the image and application of plastics packaging we can conveniently divide society and regions into groups:

- *Group 1*. Mature and often leading, for example, Western Europe, North America and Japan.
- *Group 2*. Low population densities, for example, South America, Australia and New Zealand, Eastern Europe, the former USSR.
- *Group 3*. Variously emerging societies, for example, South East Asia, Indian subcontinent and parts of Africa.

With packaging manufactured from board, metal or glass, as mentioned previously, heavy plant is required and certain skills. Consequently these are more prevalent in Group 1 countries. The presence of these packaging groups in Group 2 countries depends on population densities, what local economies will tolerate in pricing and regional export strategy. However, with quite a lot of plastics packaging manufacturing these rules do not apply; it is relatively easy in the packaging mix to manufacture plastics packaging. This allows this type of packaging to be available in all three regional groups and the restricting factors of use are the cultures of commerce and society and the economies of the country concerned.

8.4 Global trends in plastics packaging

The research programme Pack-Track™ (operated by CPS International) identifies consumer packaging innovation anywhere in consumer goods and is predictive of trends in consumer packaging. It is a qualitative study that is continuous and has been in operation for over 15 years. Specific applications of plastics packaging can be examined using this research. Because consumer packaging is so complex, topics have been grouped under four major packaging headings.

8.4.1 The application of flexibles

The application of flexible plastics packaging has been the greatest change in plastics packaging formats, or in its use, for new products launched in the past few years. It appears that there is increasing consumer willingness to buy brands and products in pouches and other flexible formats. Flexibles can deliver a credible packaged product and are relatively easy to produce.

The switch to pouches has been an important one and we see this acceptance in Europe (see Figure 8.5 of Kellogg's Muesli in a large pouch in Finland). A decade ago new brands, for example Swirls from Mars, would almost certainly have been in a carton—the traditional home of chocolates—whereas today this

Figure 8.5 Kellogg's Müsli (750 g) in freestanding pouches on the retail shelf in Finland (© Pack-Track™).

product is packaged in a glamorous freestanding pouch. There are many other examples, such as Bhalsen, the large German biscuits company; until the 1990s their products were packaged in cartons but now they are almost exclusively in freestanding pouches.

This argument can be applied to other regions. In South East Asia, Pack-Track™ has recorded a similar swing to pouches and this often includes international brands (Figure 8.6). In Japan, the application of pouches is more advanced than Europe and there has been wide acceptance of this format across nearly every consumer goods sector; it seems that if it is technically viable in a pouch, then given sensible marketing, it is commercially viable in a pouch in terms of consumer acceptance. In Australia, wine is sold in pouches (Figure 8.7). Flexible packaging, especially the pouch, is a credible, acceptable and successful consumer packaging format.

Remarkable though this trend is, it is beginning to be recognised that manufacturers are increasingly producing more of the same with this pack format.

Figure 8.6 Horlicks (as a powder) in a metallised film pouch, on the retail shelf in Malaysia. (© Pack-Track™).

Figure 8.7 Clarsac dry red wine from McWilliams, Australia, in a lively metallised pouch.

That is, freestanding pouches are all fundamentally the same thing, with differentiation only available through graphics branding and material selection; the missing element is the use of shape. This is now beginning to be addressed and Pack-Track™ is identifying an increasing number of shaped pouches on the market. These are nearly all shaped in two dimensions, although the Pack-Track™ photograph shown in Figure 8.8 is of a three-dimensional bottle in film, complete with closure.

Other examples include a face mask pack on sale in Europe, shaped snack packs in South East Asia and a shaped yoghurt pack (France, Belgium and Switzerland) where the consumer tears off the top and sucks out the contents. This is a slowly evolving area of development and production methods are lengthy (pouch making and filling) in comparison with the conventional freestanding pouch.

By switching to a pouch an environmental claim can be made as the amount of material is reduced with the additional benefit of a single substrate material. This

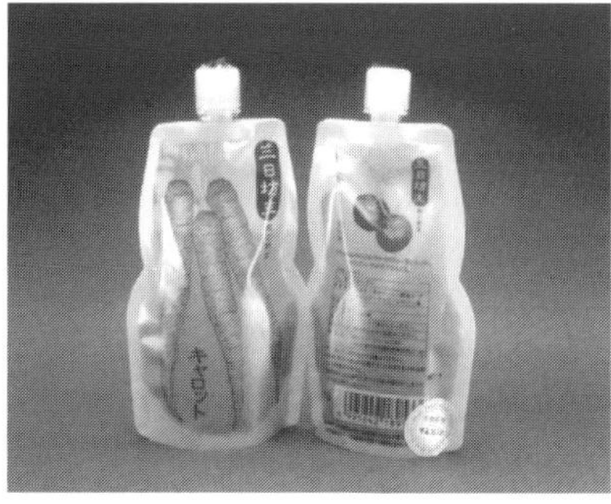

Figure 8.8 A shaped pouch mimicking the shape of a bottle from Japan. The bottle is freestanding through a gusseted base, with a pronounced waist and has a reusable screw closure to the top. Surface decoration is matt with a delicate design (© Pack-Track™).

is often a hidden statement, partly because apparently nowhere in the world is this benefit promoted to the consumer; partly because the drama of the branding through print and substrate selection overwhelms any consumer perception of environmental benefit. In fact in most global regions the environmental message, unless it is part of the brand offer, is never promoted, probably because it is a distraction from the branding or product offer.

There are examples of flexibles using an environmental platform, especially in Europe. This is particularly true in the Nordic countries and in Switzerland, where there is an incineration policy on waste disposal.

There are active areas in meeting the environment agenda with flexibles. A well established area is household laundry, an example of pioneering success for use of pouches with a technically difficult product. Another area is jams and preserves, which have shown success on a niche, but high profile, basis. Pack-Track™ has examples as diverse as Switzerland and South Africa. A more innovative approach has come from Denmark (and distributed throughout the Nordic region). The Pack-Track™ photograph (Figure 8.9) shows a retail shelf with three packs. On the right there is the original glass jar of strawberry jam, in the middle a 1 l plastics tub—this is the new container—and on the left a 'chubb' pack of jam, which is cut open and emptied into the large plastics tub. This is an example of a rare, very overt, environmental statement pack and has been successful in the latter years of the 1990s.

There is an additional area where flexibles are making inroads and that is with the inclusion of a closure or a reclose mechanism. Simple examples would include large 'bags' of breakfast cereals from Quaker, marketed in North America; these have a reseal gripper at the top. The household laundry detergents mentioned above benefit from a screw closure to open, dispense and re-close, a technique increasingly used with toiletries.

Figure 8.9 Right: the traditional (one trip) glass jar of jam. Middle: a larger plastics tub of jam; this acts as a refill durable pack. Left: a (one trip) lightweight 'Chubb' pack of jam that acts as the refill for the durable pack. From the Nordic region (© Pack-Track™).

Flexible packaging is the largest and, in many ways, the most important of the plastics packaging changes; its application is widespread and its acceptance appears secure. It represents a large part of the packaging future (and not just in plastics packaging). It needs development in shaping and improving filling speeds. It should be noted that the simple vertical form-fill bag (often known as a transwrap bag) is not covered in depth in this chapter; the text is reserved for contemporary or important packaging developments and future trends, but this simple transwrap bag format is commonplace, along with its (horizontal form-fill-seal) flow-wrap partner.

8.4.2 The growth of PET bottles

Flexibles have been an evolution, but PET has been as near a revolution as one gets in consumer packaging. At the turn of the new millennium it is showing significant signs of making serious penetration into the alcoholic, carbonated and still beverages market and early signs of success with selected milk applications. The largest market, carbonated soft drinks, has been well established for many years but this has really been at the 2 l bottle size. This container now delivers a commodity image and there are signs of introducing shape and raised surface features into this size bottle to promote brand differentiation. The real growth has been in the 1 l and under sectors. Of particular importance have been mineral waters, new age drinks, beer and the global brands such as Coca-Cola.

In this very large group there has been significant growth in highly patterned surfaces, often where the pattern is angular, evocative of ice crystals, diamonds and crystals; this is particularly common in the water market. There are lots of other shapes and designs too, such as curves and swirls. The Japanese are very fond of a spiral running down the body, sometimes reminiscent of a screw thread.

It is a strong feature of the growth that the bottles are shaped and patterned; at first this was mimicking the mood of mainly premium glass bottles. This has now evolved into far more precise and refined designs, peculiar to PET bottle technology. Coupled with this has been a general enhancement in clarity and gloss. This combined outcome is very impressive indeed. Pack-Track™ has a wide collection of quality examples from all continents, including mineral waters from Australia, an Orangina bottle from a test in France and spicy sauces from Japan. Figure 8.10 shows a typical quality example of a mineral water bottle.

This is a big, growing, plastics packaging sector and surprisingly, because of recycling issues, Pack-Track™ is identifying an increasing number of self-coloured, one-trip PET bottles, especially in a deep blue, for mineral waters most prevalent in Europe.

Use of PET for beer is more problematic, given the difficult shelf-life requirements, notably its sensitivity to oxygen, but this is now demonstrating steady growth in PET. Whereas PET for water is a single-layer structure, beer has seen the application of multilayer structures and very recently internal coatings.

Figure 8.10 Three different mineral water bottles from three different countries (Belgium, Australia and Malaysia) demonstrating simple and sophisticated surface-moulded design finish (© Pack-Track™).

Many developed countries now have PET beer bottles in full distribution. At first these were copies of a conventional beer bottle shape—this is, of course, very helpful to production lines—but as the market has strengthened, brand related shapes are beginning to emerge. The example shown in Figure 8.11 is Bier 33 from France, a multilayered bottle, although, it must be noted, there have been constant signs of layer delamination, particularly noticeable around the shoulder surface raised patterns.

An emerging sub-sector is the returnable PET bottle (as opposed to PET blends discussed in Section 8.4.3), which is often characterised by being self-coloured. This helps overcome the recycling problem and at the same time, strengthens the inherent, physical manifestations of the brand. A market in which this works well is the mineral water sector.

Figure 8.11 Bier 33 in a multilayered PET bottle. Note the raised surface design. This is a one-trip bottle (© Pack-Track™).

8.4.3 Plastics claiming to be an environmentally beneficial piece of packaging

One of the arguments promoting PET for bottles is its enhanced environmental credentials over glass. It is lighter, cheaper to transport and, above all, uses considerably less energy in its conversion into bottles and other containers. Considering the volumes used in, for example, beer, the net benefits of lightweighting/energy saving by switching from glass to PET bottles are considerable. The reduction in material in tonnage terms through plastics packaging has been remarkable and, as a consequence, an important commercial force and driver of trends. The environmental agenda drives the market in a trend towards PET bottles. PET can be shaped and is inherently lightweight (thin, yet strong), and it lends itself to actual (empty) container volume reduction at post-consumer use. This feature is achieved through structural design. In the case of PET bottles the empty bottle is collapsed, typically by pushing the bottle downwards on its base like a concertina.

Two effective applications are Evian from Europe and Charmy Compact from Japan. Charmy Compact (shown in Figure 8.12) makes a claim on the pack of weight reduction; a 40 g reduction over the durable bottle by using this (collapsible) bottle. For a 300 ml container this is a remarkable achievement.

Another approach is the so-called 'Fottle'; the word is a compaction of the expression **F**olding B**ottle**. Here the bottle has a 'V' shaped recess in the base which, when the bottle is empty, can be rolled up to aid space reduction. The material used in Fottles is often high density polyethylene. This invention comes from tiny (in population) New Zealand and is used, in one fashion or another, in all continents of the world.

The use of reusable PET and PET blends is one of the most effective applications of plastics as beneficial environmental packaging; this is just emerging in volume at the start of what could be a significant trend. Although in several countries refillable PET has been in operation for several years, for example, soft drinks including the major brands in the Netherlands, it is not widely used across Europe, nor has it been in common use in other global regions.

Beer is again the product vehicle for this move. An example is the Danish Pilsner brews from Carlsberg (see Figure 8.13) and Tubourg. Both bottles are a blend of PET and PEN, and the PEN content may be as high as 90%. The labelling on the bottle states that there is a weight reduction of 35% over a standard 330 ml bottle. The external size is the same as an original glass bottle but wall thickness of the plastics is much thinner and more consistent, which provides for an extra 50 ml of beer (approximately 15% increase in volume).

All these plastics beer bottles, one-trip and reusable, require different closures from the conventional crown cap. For Danish bottles it is a rip cap with ring pull, in Sweden a plastics screw cap with integral tamper evidence

Figure 8.12 The Charmy collapsible bottle from Japan. The wavy ribs (highlighted under the sleeve label) around the body of the bottle facilitate its collapse after use to reduce volume (© Pack-Track™).

Figure 8.13 A three-bottle multipack of Danish beer. The bottle is a PEN–PET blend and is returnable (© Pack-Track™).

and for the French Bier 33, discussed in the previous section, it is a metal plastics combination screw cap.

8.4.4 The creative urge

Plastics packaging is extremely versatile and lends itself to innovation and creativity; this makes it a potent medium in marketing and technology and therefore an equally potent force commercially. Plastics can solve many packaging problems. Pack-Track™ has a very large collection of examples of innovation, problem-solving, opportunity taking, and brand building through shape and on-pack features. It is not possible to demonstrate all these, the Pack-Track™ database is too large and diverse. One common application is to elevate the pack from a one-trip unit into a durable container offering an after-use to the consumer. This is a frequently used technique on a global scale in the spreads and soft yellow fats market. The increase in plastics weight is often minimal in justifying this move. There is also another subtle consequence and that is that these containers are often more consumer friendly then their one-trip competitors and the design more aesthetic.

Mimicking the original product shape or using shape to focus communication is a fairly common practice in Europe and can be an on-going part of the brand strategy or used on a campaign basis. For example, the large confectionery manufacture Storck used a giant Werthers Originals as a Christmas promotion. The container was a plastics replica of the individual sweet, over 10 times the size, and used as a pack/storage box, which was then subsequently wrapped as jumbo twist-wrapped sweet.

As an ongoing shape led brand, Denta Lux is a tooth gel for teeth cleaning. The pack is a squeezable container (blown bottle) with an injection moulded twist-off top. The very distinctive final shape is that of a tooth. The brand is quasi-international but may have suffered through looking more like an extracted tooth than a clean and still-in-place molar.

Other popular examples of this genre can be seen in honey packs around the world that look like beehives and some milk-based products reminiscent of traditional European milk churns, again from all around the world.

Closures represent fertile territory for innovation and Pack-Track™ has some splendid examples, the overwhelming majority of which are injection moulded. One example is shown in Figure 8.14, where the closure acts as a cleaning brush for the brand Top Pre-Care.

A flexible pouch application is a good example to consider because of the importance of this packaging format; it will also demonstrate creativity in an area where success is difficult to achieve (Figure 8.15). The closure is a channel

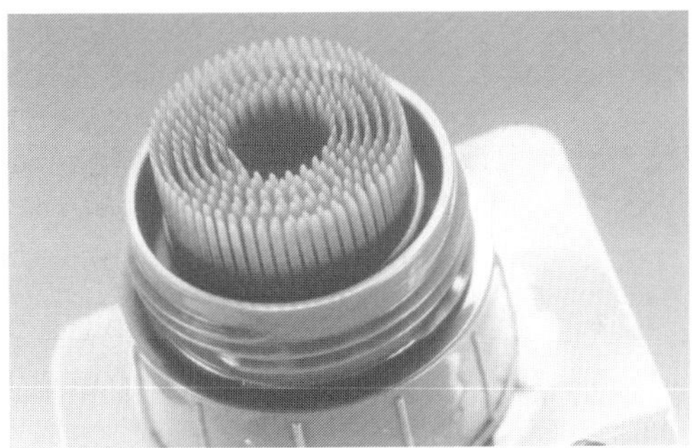

Figure 8.14 The top of a cleaning brand (Top Pre-Care) from Japan. The closure has a built-in brush to enhance the cleaning action of the liquid (© Pack-Track™).

Figure 8.15 Close-up of the top of a pouch showing an in-built channel, which allows a paste (but also good for liquids) to be squeezed out after cutting the top off. The channel is self-sealing after removal of hydraulic pressure (i.e. when user stops squeezing the pouch). The example is Japanese but the technology is North American (© Pack-Track™).

created through two sheets of flexible plastics by virtue of their not being heat sealed together. The shape of the resultant channel, the nature of the film, applied tension and internal surface treatments provide for the channel to remain closed off when the pouch is in its normal filled state. This remains true when the pack is opened by cutting the corner off the pouch, as shown in Figure 8.15. The channel only opens when the body of the pack is squeezed, applying an internal hydraulic pressure, which forces an opening at the pack's weakest point, which is this channel. The example illustrated is Japanese, where its use is popular, but the invention and technology is American. It is in use in selected liquid foods and some toiletries in the US, Canada, South East Asia, Europe and Japan.

An important feature is the extension of the closure's role into a means of controlling product delivery. Figure 8.16 shows a hair product, delivered from an aerosol as a mousse. What makes this pack special is the built-in brush, designed so that the mousse travels up through the centre of the brush and through the 'bristles', giving perfect delivery and, thus, the facility to brush in and style all without having to touch the product or the hair.

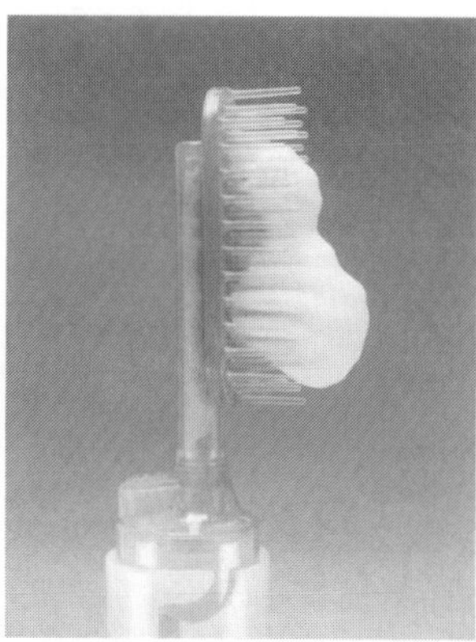

Figure 8.16 Close-up of a brush that fits on top of a conventional closure of a hair-care aerosol. The brush has a hollow channel that allows the mousse to travel up under pressure from the aerosol and come out into the brush head. From Japan (© Pack-Track™).

Extending the use of a product can be demonstrated in another way, as shown in Figure 8.17, where yoghurt packs can be interlocked to form play bricks. This creates fun in use, adds value to the brand and is a good lever to encourage repeat purchase for the consumer.

As plastic continues to expand its technological horizons in meeting consumers' needs, those needs are rapidly changing throughout the world, so it faces new challenges. In fact, there is a finely balanced debate on the perspective of consumer plastics packaging. Is market demand ahead of plastics packaging, or is plastics packaging driving the market? The answer is not too clear and naturally changes from region to region.

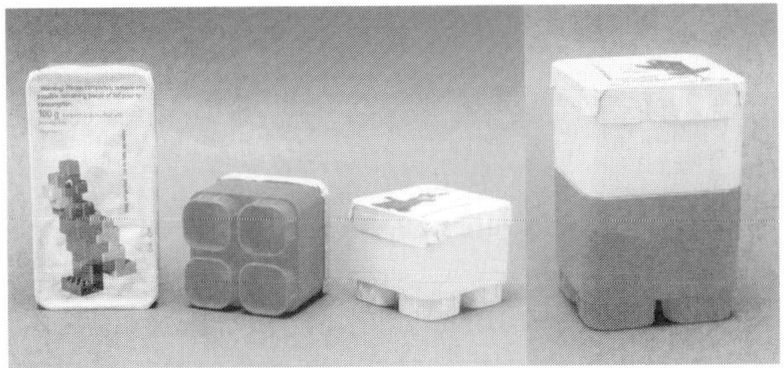

Figure 8.17 Children's yoghurt packs with a feature in the base to allow them to be used as interlocking play bricks after the yoghurt has been eaten. From Benelux and the UK (© Pack-Track™).

There are other examples. Figure 8.18 shows a bottle of liquid soap, which is shaped like a bar of soap and is claimed to float in the bath! The versatility of plastics is well developed, as seen in Figure 8.19, which shows a dispensing unit for Osyris, a household cleaner in tablet form. The wheel turns and a tablet can be dispersed without touching it. There are also child resistant and tamper evident features in the design.

In looking a little further into the future, two food-based examples are considered. The first one takes a universal commodity, cheese, and transforms it into an interactive consumer pack. The example shown in Figure 8.20 is fun, has character—literally—provides portion control, will significantly enhance the chances of the child eating the product (not an easy task!), and is wrapped with a distinctive persona from point of purchase right through to consumption. This offers powerful branding, secured through plastics packaging.

The second example is a microwaveable pouch that opens in the microwave oven when the temperature reaches a prescribed level (see Figure 8.21). This facility is provided by some clever seals between film substrates that are

Figure 8.18 A bottle of liquid soap, shaped like a traditional bar of soap that can float in the bath. From the UK (© Pack-Track™).

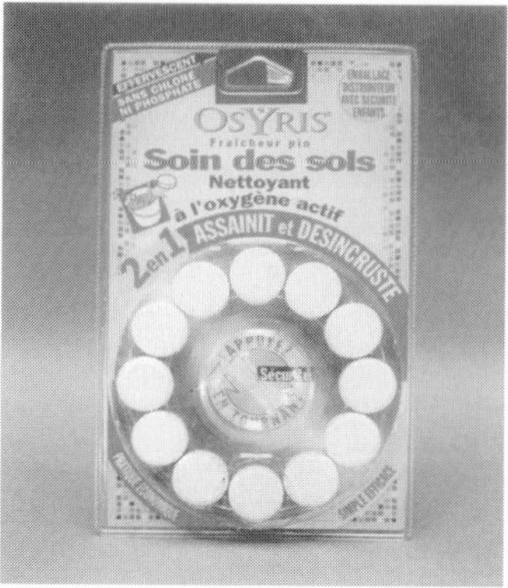

Figure 8.19 A dispenser pack for a household cleaner. The pack, which is displayed complete like this at retail, has a central circular turning device that enables a tablet to be dispensed without having to touch it. From France and Benelux (© Pack-Track™).

Figure 8.20 Little portion packs of cheese presented as characters; the top is twisted off to tear open or squeeze and eat. From the UK (© Pack-Track™).

Figure 8.21 An illustration of a pouch containing a ready meal. Three stages are identified showing the release of steam when heated in a microwave oven to leave an aperture to provide enhanced ease of opening. From Japan (© Pack-Track™).

activated by a combination of temperature levels and internal pressures created by gas and water vapour expansion on heating.

8.5 Conclusion

Society has come a long way in a short period of time with the technology that plastics packaging provides. We can buy shelf-stable foods that are safe, nutritious and wholesome. They require little or no preparation. They can be rapidly transformed into a cooked hot meal (or meal component) via a microwave oven, ready to be placed on a plate. Only the stage of actually eating out of the packaging remains to be developed and that may not be long in coming. We could quote similar achievements across a whole range of other market sectors, some of which have been highlighted.

Plastics packaging has given society a better quality of life. It has given commerce a versatile weapon in the development of trade and products, irrespective of regional size or economic status. It can be argued that plastics packaging has been supportive of the environment. These perspectives are truly international, although the actual development and application stages will be different by region.

Plastics are derived from finite oil resources, however, and the packs themselves are virtually indestructible without human intervention. Globally, millions upon millions of plastics packaging items are produced every day. Tackling in a constructive and positive way the strategic long-term use of post-consumer packaging waste is a challenge that every one of us must face. In the final analysis, it is a creative challenge, which, based on our creative history, should be achievable.

9 Plastics packaging and the environment

T. McCormack

9.1 Introduction

Populations are nowadays distributed in such a way that most of us do not live in the vicinity of resources such as foodstuffs or, indeed, possess the requisite skills to cultivate or prepare them.

Packaging provides the vehicle which enables food produce and other consumer goods to arrive at a suitable point of sale with a high level of protection and efficiency. Plastics materials have already played a significant part in the evolution of the packaging industry in the second half of the twentieth century, offering unparalleled levels of cost-effectiveness and functionality. These aspects are dealt with in more detail elsewhere in this book.

This chapter argues that plastics packaging has made a significant contribution towards sustainable development, fulfilling many of the basic and evolving needs of the community in a convenient and economic manner, while optimising resources. The effect that packaging design has on the plastics packaging market is outlined, together with the impact that plastics have on reducing the incidence of packaging as a whole.

No discussion on packaging and the environment would be complete without consideration of legislative developments and the effects that these have on the management of waste. An overview of the basic elements contained within packaging regulations is provided, along with a perspective on future trends. In addition, the approach adopted by the plastics industry towards waste management, which recommends a course based upon best available practices to meet future legislative targets, will also be covered.

9.2 Sustainable development

Humans need to consume resources in order to subsist and in doing so they have a detrimental effect on their surroundings. Accepting that we do influence our environment is the first step in considering ways in which we may minimise the adverse effects. We need to reconcile the negative impact that we have with the requirement to nourish ourselves, and fulfil other basic needs. In developed societies there are additional goals, however, such as the achievement of a certain standard of living, which may be brought about through economic growth.

It is clear that a more sophisticated and holistic model is required that encompasses not just the important aim of reducing environmental impact but also the rights of individuals and communities to a certain quality of life, choice and self-esteem and to an appropriate degree of economic prosperity. These aspects are all considered within the concept of sustainable development. Embracing this new holistic approach will place completely different demands on industry in the twenty-first century.

The concept of sustainable development is principally that we utilise the world's resources in a responsible and ethical manner, preserving them for our progeny. The most commonly accepted definition of sustainable development derives from the Brundtland Commission, and was adopted at the UN Conference on Environment & Development in 1992 (UNCED conference or Rio Earth summit):

> 'Meeting the needs of the present without compromising the ability of future generations to meet their own needs.'

The approach concerns the appropriate use of resources now and in the future; however, it does not limit itself to environmental protection, but rather widens the model to incorporate elements of economic growth and a social agenda. The concept of sustainable development may be described as being built upon three supporting columns, each of which merits consideration (Figure 9.1).

Currently, the sustainable development approach is not fully developed, and a model is required that transcends the traditional organisational measurement and reporting procedures based on financial statements for shareholders.

Industry will be expected to achieve continuous improvement in environmental performance, while striving for the complementary goals of economic added value, stakeholder satisfaction, respect for individuals and communities, and the provision of improving resource-efficient products and services. There is growing recognition that the stakeholder base is extremely broad and includes regional or country economies and society at large.

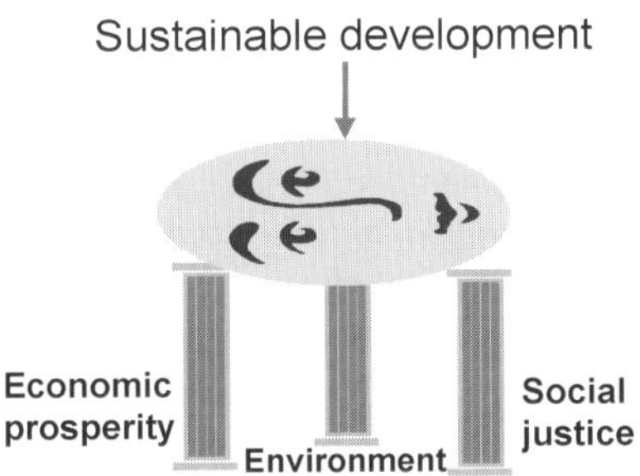

Figure 9.1 Sustainable development.

The method of reporting on this new triple bottom line model (economic added value, social indices and environmental impact) is being investigated by many industrial organisations, who recognise that they have a responsibility to integrate more of the 'softer issues' into their corporate objectives and strategies.

9.3 Plastics and sustainable development

While the plastics industry has no special insight into the complex subject of sustainable development, it is possible to consider how plastics may contribute towards those indicators of sustainable development which have been derived by the UN. Let us consider these indicators in the context of consumer goods such as packaging.

9.3.1 Economic development

The plastics manufacturing industry alone employs approximately 70 000 people, while some 1.1 million are involved in the wider industry chain. Plastics products are to be found in every major industry segment and subsegment, adding value and contributing towards wealth creation. Over 40% of all plastics currently produced are used in the dynamic packaging industry. The growth of plastics consumption outstrips GDP growth in every country, often by a factor of more than two.

Table 9.1 provides examples of UN indicators, and how they may be translated into plastics in action.

Table 9.1 UN indicators — economic, institutional

Indicator	Plastics example
Consumption changes	Packaging, consumer goods
Employment	Half million in packaging chain (est.)
Economic development	Growth *c*. 6% per annum

9.3.2 Social justice

Typical UN indicators and the plastics contribution are shown in Table 9.2. The products that people consume are not solely meeting the basic need of subsistence, but also 'make a statement about their lifestyle' and their self-worth.

Packaging is currently the prime means of product differentiation; and it is used not only to protect goods on their way to the marketplace but also to convey messages about products; it is in fact a promotion vehicle.

Plastics have become the materials of choice in packaging because of their seemingly limitless versatility. Not only are they to be found in a variety of forms from flat film to rigid containers, but they may be moulded into an

Table 9.2 UN indicators—social

Indicator	Plastics example
Population density	Packaging
Well being	Branded products
Health	Food protection and preservation, bottled water, medical packs

endless variety of shapes. Plastics are thus extremely well suited to packaging differentiation.

The present distribution of society and often hectic lifestyles demand a change in packaging to meet evolving needs. For example, as a result of families rarely dining together (the so-called 'grazing phenomenon', Figure 9.2), coupled with the rising number of single-person households, a different form of packaging is now required. This has given rise to the portion pack.

Figure 9.2 The grazing phenomenon of staggered meal times has influenced packaging.

The versatility of plastics in meeting this new packaging requirement has led to an increase in plastics packaging in those applications directly affected by this societal trend. This factor, coupled with the high degree of substitution of traditional packaging forms by plastics, has in turn caused some legislators to conclude that plastics packaging is escalating. The reality is, however, that in most cases, apart from the incidence of the portion pack, the unit weight of a given plastics package has been steadily reducing. This reduction of unit packaging weight is not merely the result of environmental altruism but also a response to the need to provide cost-effective packaging solutions.

Plastics packaging contributes in a resource-efficient way towards the protection of our foodstuffs, delivering them safely and hygienically into our homes. In the same way medical products may be mass-produced and distributed widely in a cost-effective manner, making a major contribution towards health care.

9.3.3 Environmental protection

UN indicators for the environment are shown in Table 9.3. The inherent light weight of packaging in plastics contributes greatly towards environmental protection by reducing the consumption of resources.

Table 9.3 UN indicators—environmental

Indicator	Plastics example
Sustainable agriculture	Agricultural film
Atmospheric protection	Reduced emissions
Resource preservation/extension	Lightweight packaging, high leverage
Waste management	Recycling and recovery

Approximately 4% of the oil consumed in the world goes into the manufacture of all plastics and in turn roughly 40% of the plastics currently consumed in Europe is used in packaging applications. Figure 9.3 shows the amount of oil used to produce plastics.

In 2000 plastics packaging accounted for about 19% by weight of all packs and yet plastics are used to package half of the products—this is phenomenal leverage.

Over the past ten years, the plastics packaging industry has reduced the amount of product used in an average application by 28%, and this is ignoring the inter-material substitution trend that has occurred and will continue to be a major factor in the future.

There is no reason to assume that the source reduction values that have been attained will not continue to occur as a result of material property developments, inter-polymer substitution and packaging concept change (e.g. movement from bottle to flexible pouch).

The importance of source reduction, should be noted, even though light-weighting, through component wall-thickness reduction or via concept change, may actually have a detrimental effect on mechanical recycling. Owing to established recovery systems for traditional materials this method is often regarded as the only means of dealing with waste.

Plastics suffer from comparisons with other packaging media, which lend themselves more readily to mechanical recycling. This problem arises from the

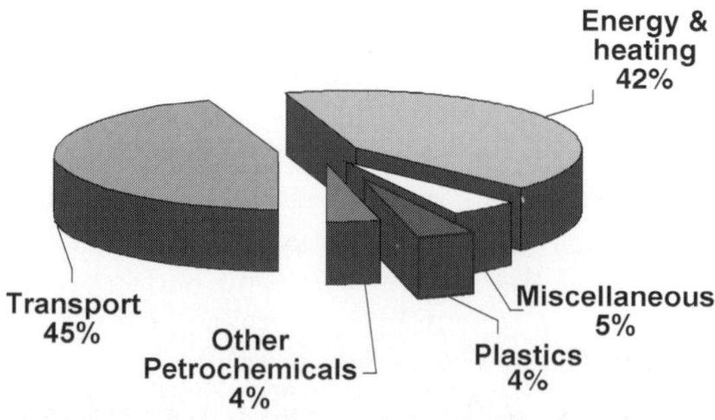

Figure 9.3 Application of petroleum products in Western Europe. Only 4% of commercially produced oil is used in plastics production; 87% of oil goes towards transport, heating and energy. Source: IEA APME 98.

misapprehension that 'plastics' are one material, when in reality, as the name implies, they are a category of materials with widely differing properties and behaviours.

9.4 Packaging as a concept

Packaging is the most cost-effective means of conveying goods to the market-place. It serves a key role in reducing spoilage of foodstuffs and therefore saves valuable resources and energy. One needs only to compare the low 2% wastage in a mature or developed region with that of the developing regions of the world, which currently lose an estimated 40% of their produce as a result of spoilage.

9.5 Leverage

The use of plastic materials is growing because they are currently the materials of choice in almost every industrial segment, bringing as they do wide performance windows and property profiles and effective use of resources.

As this book seeks to demonstrate, it is the sheer variety of plastics types which enables a broad range of packaging applications to be attempted. It is important to consider the entire life-span of plastics in packaging and to recognise their advantages and their contribution towards our modern lifestyle and not merely to focus on the waste management issue. This latter point is no easy task because the eye-catching tactile pack, which we select in the supermarket to transport our favourite foodstuff safely home, soon metamorphoses into a bulky container that needs to be squashed into the disposal receptacle (Figure 9.4).

Plastics packaging materials typically pack products hundreds of times their own weight; an example of the plastics packaging leverage for transport packaging is depicted in Figure 9.5.

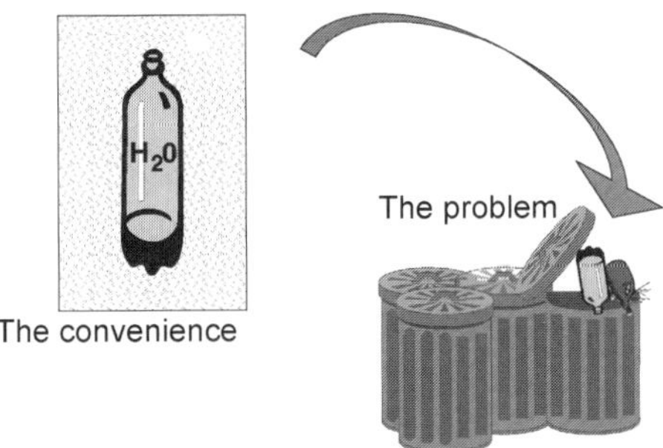

Figure 9.4 Convenient packaging still presents a disposal problem.

- 1375 kilos packaged
- Bag contents 25 kilos
- 55 bags with each unit weighing 110 grammes
- Over-wrap 1.7 kilos
- Total plastics packaging weight of 7.75 kilos
- Weight of packaging per tonne packed is 5.6 kilos
- Leverage ratio of 177:1

Figure 9.5 Plastics packaging leverage potential.

So why are there so many plastics types in the packaging sector? The answer, as highlighted elsewhere in this book, is that plastics materials provide distinct functionality for different applications. They are also used in combination, where the synergistic effects between various materials provides the correct degree of overall protection for a given product. This tailoring of properties such as water and oxygen barrier, stiffness, impact, clarity and sealing characteristics means that many types of plastics with potentially widely different chemical compositions are employed in the packaging sector.

9.6 The anatomy of a dustbin

The contents of the average household waste across Europe should be considered to ascertain the real magnitude of the plastics problem and place it into some form of perspective. The actual proportions across Member States differ as many countries now operate segregated schemes for household waste and some traditional packaging media such as glass have recovery systems which are long established. Municipal solid waste (MSW) contains: paper, vegetable matter, textiles, metals, glass, wood and plastics (which in 1998 accounted for 9% of the MSW by weight).

Overall, the packaging within a typical dustbin is the same as it was 20 years ago; that is, 2–3 kilos are discarded per week (source INCPEN). Over this time-frame packaged products have undoubtedly increased; however, the unit packaging weight has decreased. Some 25–30% of the total weight of MSW is due to packaging. Considering the packaging portion only in a dustbin in 2000, the plastics fraction is estimated at 19% (see Figure 9.6).

9.7 The situation without plastics packaging

As we have grown up with plastics it is difficult to imagine life without them. However, a recent German study considered the implications of a return to traditional packaging media. The GVM study concluded that the net impact on

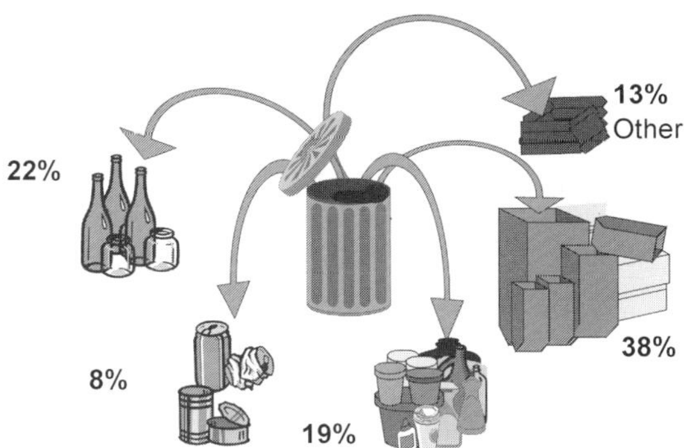

Figure 9.6 Packaging contents of a dustbin in 2000.

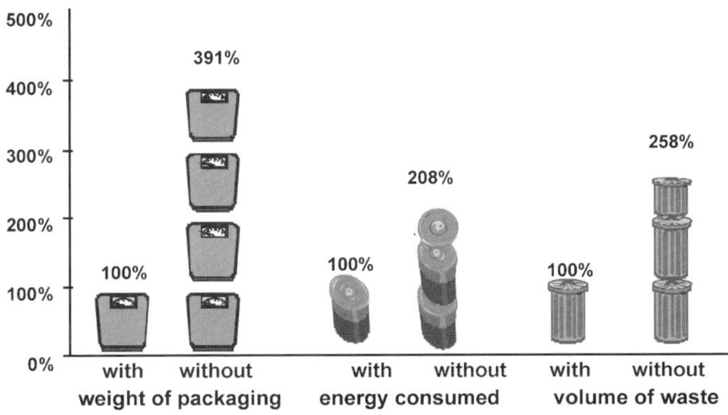

Figure 9.7 Packaging *without plastics*—using alternative solutions. Source: Gesellschaft für Verpackungsmarktforschung (GVM).

the environment in terms of weight and volume of packaging to be dealt with would increase significantly and that, in addition, the energy consumed would double if alternative packaging forms to plastics were to be adopted. The impact of life without plastics in the packaging industry is summarised in Figure 9.7.

9.8 Waste management (WM)

Plastics are not a single homogeneous category of material and there is a wide variety of WM options available for plastics compared to more traditional packaging materials such as paper or glass. Plastics materials are a victim of their own success as they provide an optimal service in terms of cost-performance bearing in mind the variety of jobs they do, and in addition they are extremely lightweight. However, the value of the recyclate is correspondingly low.

9.8.1 Prevention

This is a very important element in waste management that should not be underestimated, because its prime purpose is to reduce the magnitude of the task in hand. As mentioned earlier, a conservative estimate is that plastics packaging has been reduced by 28% in weight over the past decade in existing applications. The following sections consider the impact that this has on the generation of waste.

9.8.2 Preventative design

Packaging designers are often not given enough credit for reducing the amount of packaging in the marketplace. In addition, legislators, despite including prevention as a key objective, do not always recognise the impact that it has on packaging. Plastics in particular have already made a significant contribution towards packaging prevention. The number of examples of packaging prevention in the plastics sector are legion and may arise through:

(i) development of new materials;
(ii) optimal component design enabling grammes to be shaved-off the weight of rigid components;
(iii) tougher films which may be down-gauged without loss of properties;
(iv) technology switch (e.g. from a thermoform to an injection moulding with better control of wall thickness);
(v) change from pre-made pots to form-fill and seal container systems; and
(vi) substituting a dense for a less dense polymer, for example, structural foams, which can provide significant weight reductions over compact plastics.

Another key element in reducing component weight is switching to an alternative polymer family, and this intra-material substitution within plastics often occurs as packaging is rationalised or harmonised. However, the packaging designer of today needs to be well versed in the full inter-material substitution (IMS) opportunities, and here, plastics have a very important role to play. Programmes to decrease packaging weight are very much in focus today, owing to the added incentive of reduced Eco-taxes.

9.8.3 Technological innovation

Inter-material substitution based upon plastics becomes more likely as technical innovation continues to improve plastics properties, ease of processing and economics. There is a creative evolution arising from the development of new materials and improvements in conversion technology. For example, the improving stiffness/impact balance for polypropylene results in improved top load without loss of toughness, and consequently permits components with thinner walls to be produced.

9.9 Packaging trends

Within the dynamics of the packaging market opposing forces have a profound effect upon plastics consumption and the amount of packaging waste that is consequently being generated. Source reduction and concept change, which are continually occurring as a result of innovation in plastics raw materials and evolution of packaging design, are driving the unit weight of existing plastics packaging applications downwards.

The natural underlying market growth of packaging, as consumption expands, is increasing the demand for existing plastics packages. In addition the property profile and cost-effectiveness of plastics raw materials make them the materials of choice when confronting the increasingly sophisticated demands of the modern packaging industry. Thus, the versatility of plastics in meeting the evolving societal needs is fuelling their growth in packaging, in substitution of alternative packaging media.

Furthermore, the changing consumption trends arising from the new societal scenarios, such as the grazing phenomenon, are causing an effective increase in plastics packaging in some established applications. This subdivision of the packaging unit results in the introduction of portion packs. It is worth remembering, however, that the prime aim of this packaging evolution is in response to the over-riding requirement to reduce product waste, which in turn conserves resources and energy. Consider the fact that 1 kg of meat requires 63 MJ of energy to produce, while the production of a plastics packaging film, to ensure its protection, utilises a mere 2% of this energy. The packaging market trends are summarised in Figure 9.8.

The overall contribution that plastics makes to packaging may be determined by assessing the impact of not having this packaging form available. As discussed earlier, the study by GVM indicated that a substitution factor of 3.9 by weight is appropriate when one considers the relationship between plastics and previously used packaging materials. However, as shown in Figure 9.8, there are additional factors to consider. The actual unit packaging weight of plastics

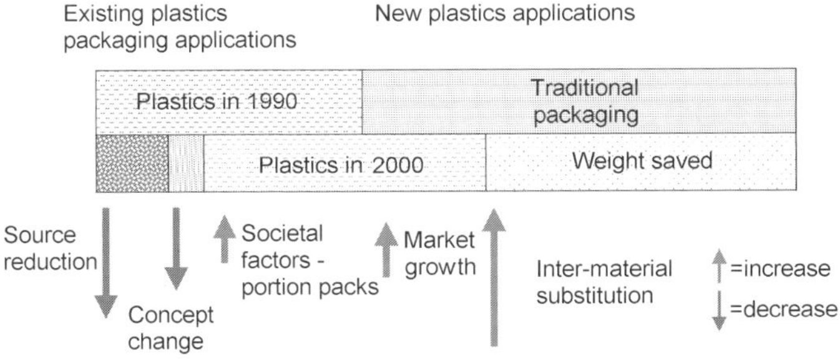

Figure 9.8 Plastics packaging trends.

materials has been decreasing as a result of two factors:

1. source reduction, where material improvements enable the same performance to be achieved at lower component weight, and
2. switching from one design concept to another, which permits drastic reduction in the resulting packaging weight.

On the other hand, the consumption of plastics in packaging is increasing as a result of underlying market growth, unit packaging breakdown (portion packs) and through the substitution of other packaging forms.

The portion pack effect, due to societal change, has been estimated as being responsible for one third of the growth of packaging in general, in France, from 1994 to 1997. For the purposes of this study, it is assumed that this is a continuous trend and that the French model is representative of the total scenario across Europe. These assumptions are made to enable the potential magnitude of the market changes to be assessed. A more in-depth, rigorous study would be required to quantify the numbers more precisely.

The contribution of all of these elements needs to be considered when calculating the overall impact of plastics packaging on today's market.

9.10 Impact of plastics packaging

The actual contribution made is considered in this section. Using the following formula it is possible to estimate the substitution effect that has occurred over the past ten years.

Consumption today (PP2000) is made up of the following elements:

- existing plastics packaging applications (PP1990),
- proportion of plastics that has undergone concept change (CC), resulting in a dramatic decrease in weight,
- the proportion that has undergone concept change to meet the societal evolution (SCC), i.e. the demand for smaller portions and the subsequent increase that has had on packaging weight per unit product packed,
- proportion of applications that have light-weighted (SR),
- the underlying growth in the packaging market, and
- the differential plastics growth, resulting from substitution of alternative packaging forms and the preferential selection of plastics in new market creation (IMS).

Figures 9.9, 9.10 and 9.11 show the dynamics of the packaging industry. Estimates are given for the net plastics contribution over the period 1990 to 2000 based upon a variety of European studies. (The full range of assumptions is summarised in Section 9.11.1).

The figures clearly show that inter-material substitution has had a very profound effect on growth in plastics volume. This stark conclusion may lead to

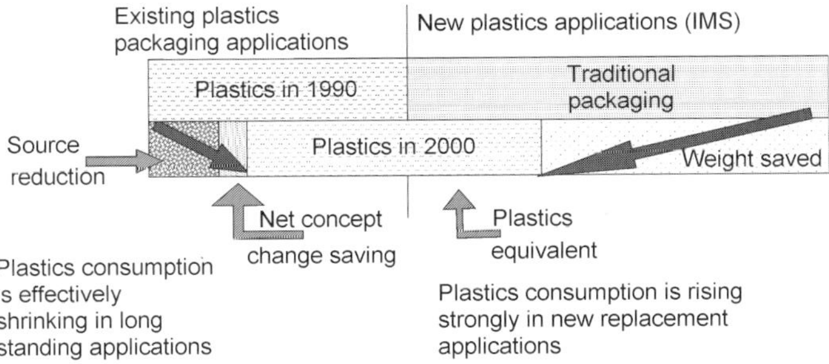

Figure 9.9 Plastics contribution to resource savings/optimisation over the past decade.

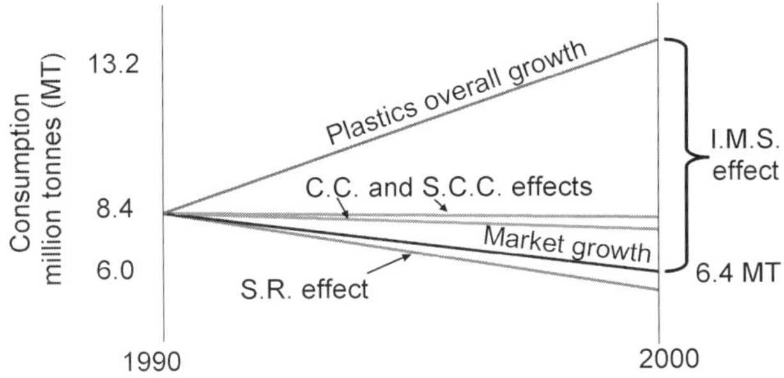

Figure 9.10 Plastics contribution to resource savings/optimisation. S.R. = source reduction; C.C. = concept change; S.C.C. = societal concept change; I.M.S. = inter-material substitution.

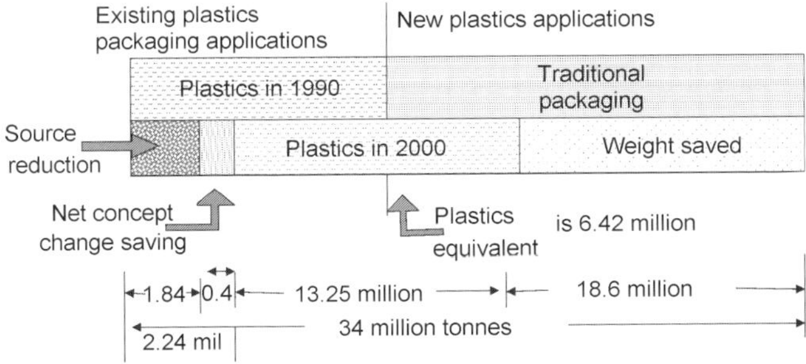

Figure 9.11 Plastics contribution to resource savings/optimisation (2000 estimate; see Section 9.11.1). Total saving equivalent to 20.8 million tonnes p.a.

the mistaken conclusion that plastics materials are not taking due consideration of the growing waste issue. In fact the contrary is true. When actual values are assigned to the plastics packaging scenario it becomes apparent that the source reduction and net concept change have a strongly positive effect on weight reduction and result in 2.24 million tonnes of plastic being saved.

9.11 Total resource savings in packaging through plastics use

This calculation has been based upon the assumptions around percentage concept change and accepts that not all of the advances in alternative packaging materials, which have succumbed to plastics replacement in recent years, have been considered. The total contribution may be calculated based upon the 3.9:1 substitution ratio (derived from the GVM study). The net result is that packaging waste currently estimated to be 69 million tonnes across Europe would rise by a staggering 88% to 129 million tonnes.

Debate will continue about whether the materials being replaced by plastics today, due to the (albeit indirect) consumer demand, are more sustainable. The fact remains that plastics are saving us a huge packaging burden, consume relatively few resources in their manufacture and use, and offer a wide range of waste management options. (An overview of the waste management options available to plastics is provided in Section 9.12).

Before leaving this treatise on how the world of waste might look in the absence of plastics it is worth considering that a return to original packaging forms is not a realistic option. The degree of protection and hygiene achieved today with the extensively used plastics packaging, coupled with the quality and the cost-performance that is demanded, would not permit the change in the majority of cases.

9.11.1 Assumptions

The numbers shown are based upon the following detailed assumptions:

(i) 10% of the market has undergone concept change (e.g. bottle to refill pouch) with a consequent 80% reduction in packaging weight (estimates);
(ii) the effect of societal changes (i.e. smaller packs are pushing up the plastics consumption in some existing applications). One third of the increase in consumption over the past ten years is due to this effect (French study) and it results in an average 30% weight increase in existing applications;
(iii) plastics packaging has reduced on average by 28% over the past decade (Association of Plastics Manufacturers in Europe, APME);
(iv) the plastics packaging market in 1990 was 8.4 million tonnes, and that the overall packaging market sustains an average 1.5% per annum growth (various TNO–Sofres studies for APME);

(v) the plastics substitution weight factor is 1:3.9 in relation to traditional packaging media (GVM study).

9.12 Plastics waste management

The contribution plastics make to the prevention of packaging waste have been considered. Methods for disposal of plastics components at the end of their useful life are now examined.

9.12.1 Reuse

This applies to articles such as glass deposit bottles. There are also some good examples in the case of plastics, mostly within the durable packaging sector, where this works well (e.g. crates, pallets, corrugated board, foldable boxes). Effective reuse, however, should consider factors such as: the environmental gain, availability of an infra-structure, and general product stewardship principles such as effective cleaning processes, safety and hygiene.

The formula to consider is: production of a new component from virgin plastic **less** collection, sorting, cleaning, refurbishing, and reintroduction impacts **provides** positive net environmental gain on the basis of a single trip.

The reuse loop should also be based on economic viability, and this is usually dependent upon the number of cycles or trips that the components can make during their service life.

9.12.2 Recycling of plastics

This also covers feedstock recycling; however, mechanical recycling is examined in more detail here.

9.12.2.1 Mechanical recycling

As mentioned earlier in this chapter there is a widespread belief that recycling is in some way the most acceptable form of waste management and that the higher the level, the better for the environment. This is in fact untrue for many materials, including plastics. The problem chiefly lies in the ecological impact resulting from the extraction, segregation, and treatment of the waste and the choice of which streams to target. Plastics add further dimensions, namely:

(i) as a result of their high leverage, they are very lightweight compared to the goods being packed, which results in a high level of product contamination;
(ii) the high efficiency of plastics packaging in the first place means that their intrinsic economic value is low; and
(iii) plastics are extensively used in food packaging, which substantially limits its recyclate being employed in its original applications. Furthermore,

collection operations may not be entered into without due consideration of potential health issues.

So with regard to the plastics packaging industry, where are the best sources of products found that should be recycled mechanically in this way?

The approach to be adopted is one of selectivity where only those plastics with the requisite profiles are selected. The best source is not the MSW at all, because of the high attendant cost in extracting quality product of a requisite consistency and in sufficient quantities to make it viable. The best source is, in fact, industrial scrap, and secondary or tertiary packaging from large institutions and trade outlets.

9.12.2.2 Plastics loops

The recycling (and recovery) of plastics may be considered as a series of loops, all of which concern mechanical recycling.

The closed loop. This applies to those forms of durable or heavy section packaging, recycled back into their original application.

Extended loop. This refers to articles, which are recycled into a different segment or in the case of packaging an alternative application within the segment.

Examples include the conversion of one-way packaging into durable products; for example, polypropylene tubs collected from segregated municipal solid waste can be converted into pallets.

The formula to follow when gauging whether a plastics component should be considered for recycling is: recyclate demand is equivalent to technical acceptance multiplied by market acceptance. For example, where technical acceptability includes food approval then demand becomes zero.

9.12.2.3 Design for recycling

It is incumbent upon the packaging industry to consider ways of ensuring that those streams which are suitable for mechanical recycling are not hindered by poor component design. For example, a bottle that leaves a tear strip of a different material is non-optimal; a tamper evident band which breaks away with the closure would be more appropriate.

Subscribing to the mono-material concept when appropriate, applying suitable identification marks and constructing articles so that mixed materials may be easily separated are minimum requirements for the packaging designer.

9.12.2.4 Recycling in perspective

A recent report commissioned for APME forecasts that the demand for mechanically recycled plastics emanating from the packaging industry will be 15.4% by 2006. One large section of the market will not tolerate the use of recycled plastics and that is the food industry. This is a severely limiting factor as it is estimated that 60% of plastics packages are utilised in the food sector.

There is also the question of specifications and the correct fit between plastics demand and availability of single type plastics. Mixed plastics, which makes up the majority of plastics packaging waste that is 'segregated' today, has significantly reduced scope for application as a recyclate.

While the conversion of a short-life article into a durable product may be a valid means of mechanically recycling some plastics packaging, the number of applications will quickly become saturated. This is particularly the case as advancing legislation has two effects:

(a) member states that currently import mechanical recyclate satisfy the demand from within their own infrastructure; and
(b) target recycling quotas covering the durable sector are introduced and the intra-sectoral transfer of plastics is invalidated (for example, automotive manufacturers would prefer to source material from within their own segment).

9.12.2.5 Open loop systems

These may be described as relating to non-mechanical recycling. However, this category does include other forms of material recycling, namely feedstock and chemical, along with other thermal recovery methods based on the combustion of segregated and mixed waste, the use of plastics as fuel or as a reducing agent. In addition there is composting, which is limited in the case of plastics.

The very high calorific value of plastics means that they make a substantial energy contribution when employed in combustion processes. In fact the removal of plastics from the MSW necessitates the addition of oil or gaseous hydrocarbons (that have not gone through an additional useful life), which are often reduced in effectiveness in comparison to the well-distributed plastics packaging fraction (e.g. small packaging films). Owing to the displacement of petroleum derivatives the use of plastics waste for thermal recovery (heat generation) may to a large extent be considered CO_2 neutral.

9.12.2.6 Landfill

It cannot be stressed enough that this is regarded as the least defensible waste management option. It is the equivalent of an open-ended loop, whereby natural resources are effectively buried. Maximum diversion from landfill is the target of everyone in the plastics industry.

9.12.3 Eco-efficiency model

The Eco-efficiency model commissioned by APME indicates that the best waste management scenario is based upon 15% recycling (mechanical and feedstock) and 85% energy recovery. The study indicates that even if recycling is increased from a 15 to 50% rate, the environmental gain is negligible and the costs increase by a factor of three. Furthermore, the best environmental/economical balance is achieved if industrial packaging is targeted and not MSW. Figure 9.12 shows the typical approach when considering a balance of potential waste management options; the target zone in the top right hand corner. Figure 9.13 indicates the ideal scenario and that achieving a 50/50 split between energy recovery and recycling provides limited ecological improvement at higher cost.

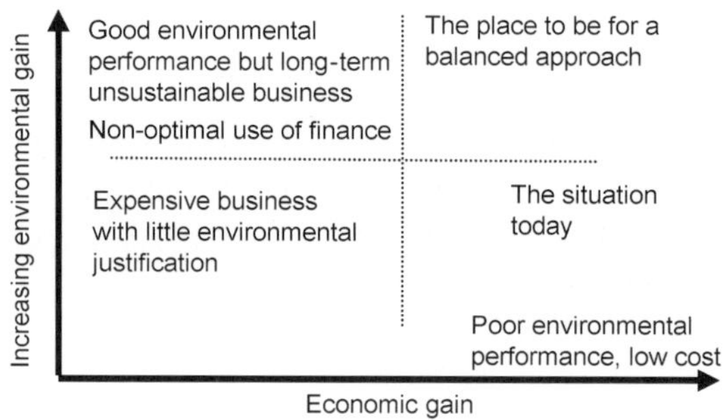

Figure 9.12 Eco-efficiency model for a balance of potential waste management options.

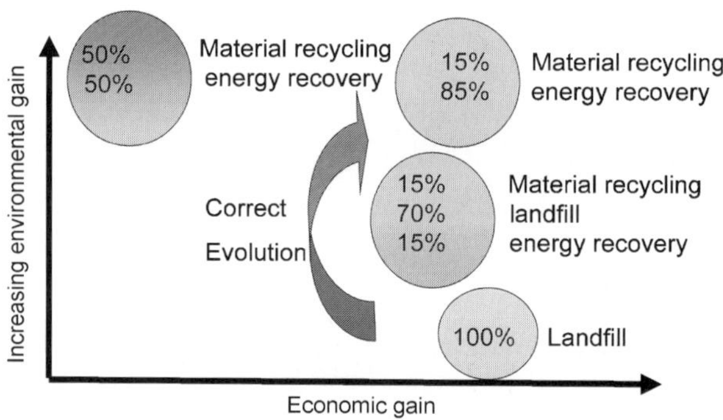

Figure 9.13 Eco-efficiency model showing relationships between energy recovery and recycling (APME).

9.12.4 Waste management options

It is important to remember that all of the above mentioned waste management options (WMO) are valid depending on the context, with the exception of landfill, which is regarded as a true waste. The objective of the plastics packaging chain is to achieve a maximum diversion from landfill. To draw an analogy, the landfilling of plastics components may be regarded as being the equivalent of pouring oil (albeit in a stabilised, solid form) into the ground!

Once landfill is set aside from the waste management equation then all of the other recovery methods have validity depending upon the prevailing circumstances. The objective should always be environmental gain and then the options may be selected on merit.

Several factors should be borne in mind when assessing the best waste management option:

(i) the facilities which are available;
(ii) the real cost of the operation (gate fee), discounting artificial subsidies; and
(iii) the overall environmental gain of following the selected route.

Clearly if the impact on the environment when the rolled-up inputs and outputs relating to the collecting, sorting, cleaning, preparing and reprocessing operations is greater than when energy recovery is employed, then the mechanical recycling route may be regarded as suspect.

Where potential routes are deemed broadly equivalent in terms of economic and social impact, then the approach that provides the greatest environmental gain should be selected. It is worth remembering that not all mechanical recycling of plastics is practicable, or good for the environment.

9.13 Life-cycle analyses (LCA)

LCA can provide a valuable tool for the generation of appropriate data for assessment of the environmental impacts of differing packaging systems that deliver an identical service. An LCA is a cradle-to-grave approach, which aims to assemble the inputs and outputs of a given system (Figures 9.14a and b).

To assist the generation of data on plastics systems, APME has consolidated the data for all of the main polymer families to form life-cycle inventories or Eco-profiles. The inventories can be used by plastics suppliers to consider where improvements can be made in the production of polymeric materials.

9.14 Legislation

The legislative position for waste management is now considered, as is the plastics industry approach to such legislation.

9.14.1 The Packaging and Packaging Waste Directive (P&PWD)

No treatise on plastics and the environment would be complete without consideration of the legislation affecting this market today. The European Packaging

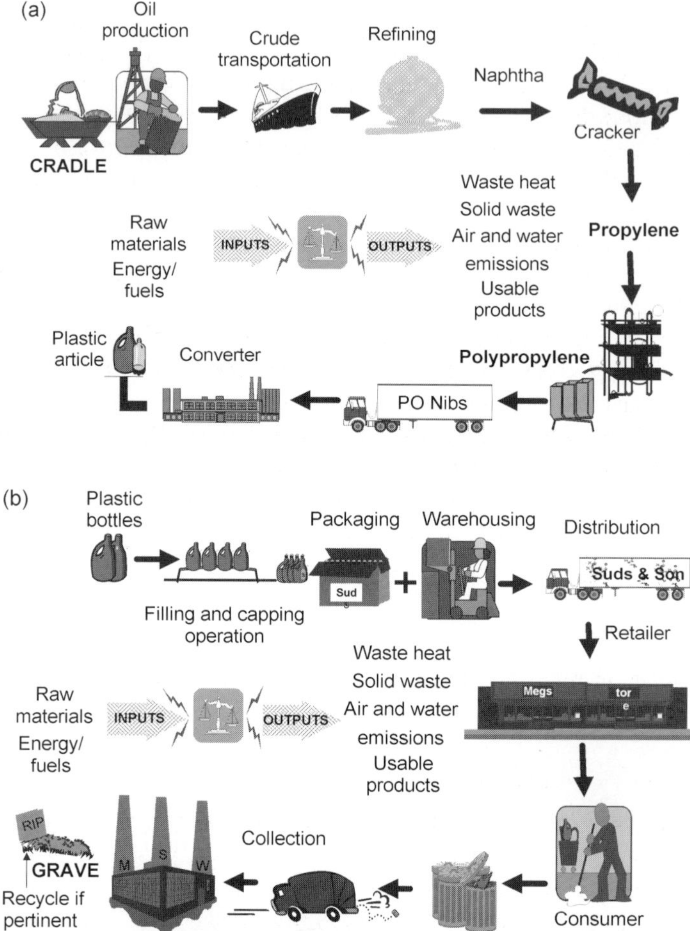

Figure 9.14 Elements of an eco-balance: (a) Production of plastics articles; (b) Service life and waste management.

and Packaging Waste Directive (P&PWD) will be taken as an example. Enacted in 1994, the P&PWD provides the framework for all Member States to address the packaging market and to deal with packaging waste. Targets and timings are laid down, although not all of the Member States are able to adopt the measures in the prescribed time-frame and certain provisions have been made for those countries.

The Directive aims to harmonise national measures concerning the management of packaging and packaging waste, in order to provide a high level of environmental protection by reducing the impact of packaging, and to ensure the functioning of the internal market, by avoiding trade obstacles and restriction of competition. The Directive provides a definition of packaging and goes on to establish the main requirements for packaging materials. Prevention is mentioned as being a key element in reducing the packaging waste burden.

The Directive decrees that within five years the Member States should achieve between 50 and 65% recovery of packaging waste and between 25 and 45% recycling of the total amount of packaging contained in the waste; with a minimum recycling level of 15% by weight of each packaging material. In summary, therefore, with regard to plastics packaging, a minimum of 50% recovery and a minimum of 15% material recycling should be achieved by mid-2001.

The targets to be achieved for the subsequent five-year period (i.e. by 2006) are currently up for review. Discussions on the target revision are ongoing but various scenarios have been proposed which could have a significant impact on plastics. These include: a substantial increase in both recovery and recycle targets; the possibility that recovery will be removed, leaving only a recycle minimum requirement; that sub-quotas for packaging materials could be imposed; and that feedstock recycling may be reclassified as a recovery method.

9.14.2 *Marking*

The commission must lay down a classification and identification system for materials, which will facilitate recovery.

9.14.3 *Essential requirements*

By January 1998, packaging allowed on the market should conform to certain 'essential requirements', which include the minimisation of packaging weight, content of heavy metals and noxious and hazardous substances. In addition, it provides for an assessment of the suitability of the packaging for a range of recovery methods.

In order to facilitate this part of the decree the Commission mandated CEN, the European Committee for Standardisation, to prepare standards. In all, six are in preparation, an umbrella standard which may be regarded as a guide to the methodology and explains their application and five standards on prevention, re-use, material recovery, energy recovery and organic recovery. These standards are currently under review.

9.14.4 *Legislation and the plastics industry approach*

9.14.4.1 *Common global objectives*

The fundamental objectives of reducing the packaging waste burden, while not compromising the working of the internal market within the EU Community, are fully endorsed by the plastics industry. However, the plastics industry does not identify with a waste hierarchy in which one recovery option is preferred to another, but would rather promote a more holistic approach which starts with the basic premise that landfill is a true waste of resources and should be avoided. In addition the incineration of waste containing plastics without recourse to energy recovery is also considered reckless.

9.14.4.2 *Targets and their revision*

The plastics industry believes that any increase in the recycling and recovery targets should be based upon sound environmental and economic rationale; and that eco-efficiency studies or cost/benefit analyses should be a key consideration when determining future objectives.

There is widespread concern that the plastics recycling target under discussion could outstrip the facilities available to achieve such a volume. Furthermore, there are insufficient end-use markets for the recyclate and an imbalance between the plastics recovered and the market demand.

As mentioned earlier, the APME sponsored report indicated that by 2006 the industry could recover a maximum of 15.4% of plastics via material recycling and that the remainder needs to be dealt with in an alternative manner. In fact the mechanical recycling of plastics packaging has always been done on a selected basis, where only those packages that can be collected and separated in an economically viable way are pursued.

Recovery should remain as a valid waste management category, since the various options contribute significantly to the achievement of the overall objectives laid down in the Directive.

Furthermore, due care and attention should be given to the consolidation of the statistics when comparing the respective performances of Member States, and current achievements should not become the basis for establishing future targets.

The optimal mix of recycling and recovery should be determined by the individual Member States, based upon the locally prevailing conditions (market, investment cycles, economic and social factors).

All of the packaging material organisations are opposed to the introduction of sub-quotas. Provided packaging materials conform to the essential requirements, then differentiation could lead to a distortion of competition. Furthermore, in the absence of evidence which demonstrates clear environmental improvement such differentiation is unjustifiable.

9.14.4.3 *Marking*

Identification of the material types is an important factor in achieving improved separation and segregation, and easing the job of the recycler. This should be coupled with the use of compatible materials for the constituent parts of a given article.

With regard to marking of plastics packaging, the Association of Plastics Manufacturers in Europe (APME) recommends and encourages the use of the SPI identification symbols.

9.14.4.4 Standards

The plastics industry wholly endorses the adoption of an agreed set of standards by which the conformity of packaging may be measured.

9.14.4.5 Definitions

The main issues for the plastics industry arise from moves to change the definitions within the original Directive and in related Directives concerning waste.

The plastics industry is strongly opposed to attempts to reclassify feedstock recycling as a recovery technique and to introduce a minimum or threshold calorific value for waste streams. The latter has been born out of the need to reduce cross-border trade in waste. However, it may serve to exclude some important processes from counting towards the targets by classifying them as disposal via incineration instead of energy recovery.

Incineration without energy recovery is considered suboptimal as thermal recovery can go a long way to realising the potential within the plastics, but it too suffers from problems, the most significant being the determination of appropriate sites.

9.14.4.6 Statistics

When considering recycling and recovery performance it is worth noting that the Member States have achieved widely differing levels in their pursuit of the common targets. The average energy recovery level is 22%; however, Denmark cites a 75% figure. With regard to material recycling, the average is 17% but half of the countries have achieved less than 10%.

The situation is further compounded by the non-homogeneity in the gathering and reporting of the statistics, such that cross-country comparisons are difficult. In addition, differing recovery technologies exist across the Member States.

As a consequence, every country within the European Union (EU) has its own approach, with regard both to achieving the targets and funding the operations. The cost of plastics packaging therefore varies from country to country. In fact we have one Directive and 15 different approaches at the Member States level.

Some countries have an entire infrastructure, to fund and deal with the waste issue, which obscures the true cost of meeting the targets. The materials associations do not support a lack of transparency and artificial funding, as it is not an economically sustainable situation.

9.14.4.7 Essential requirements

The plastics industry is preparing a guide, under the umbrella of APME, to assist with the conformity assessment, which is largely the responsibility of the packer-fillers. This guide serves to indicate where relevant information relating to heavy metals and hazardous constituents may be obtained, and provides, for example, formulae for calculating calorific gain.

Remarkably, even a plastics container produced from a material which is heavily filled with an inorganic material (e.g. polypropylene loaded with 50% calcium carbonate) will still generate a strongly positive calorific gain. In fact few plastics combinations have a negative calorific gain (unless combined with traditional materials). These evaluations support the energetic argument for the adoption of thermal recovery as a valid waste management route.

9.15 Conclusion

This chapter has sought to place plastics packaging into perspective when considering the overall consumption of materials in this important market segment. It is worth noting that total plastics represent only a very small fraction (less than 1% by weight) of the overall waste arising in Europe.

Plastics will continue to evolve and remain a major material of choice due to cost-performance and because they meet the changing and increasingly demanding specifications of the dynamic packaging sector. LCA studies will enable plastics packaging to optimise performance continually in terms of minimal impact on the environment.

Improved design and innovative product development should be foremost in the minds of the packaging chain, as this form of prevention will reduce the overall magnitude of packaging waste. Source reduction will continue to be an integral feature of the plastics packaging industry. Continuous improvement will also be demanded in the manufacture and recovery of plastics components, through re-use and recycling where it makes environmental, practical and economic sense and via energy recovery to realise their latent potential. Polyolefins, which dominate the plastics packaging market today, are expected to continue to consolidate their position, through property enhancement.

Synergistic combinations of materials should be employed to maximise functionality, although the disposal of these articles will be a prime consideration at the design and development stages. Packaging legislation with the aim of reducing the amount of resources that are consumed will continue to penalise landfill, which everyone involved in the packaging chain recognises as unsustainable in the longer term.

The pressure on the plastics industry with regard to packaging waste management is not expected to abate and needs to be fully integrated into development and business processes.

The widespread use of plastics actually serves to extend our natural resources. The versatility and wide range of materials families mean that there are in turn a variety of WMO available for dealing with end-of-life plastics. The discrepancy between the available plastic wastes and the market demand, coupled with poor economics and questionable environmental gain, will always restrict the amount of plastics that are mechanically recycled.

Augmenting the recovery of plastics from the household waste stream for material recycling will become more difficult to attain as a result of:

- the limitation of the resources to deal with it effectively
- the fact that the economics of progressively recovering plastics becomes increasingly unfavourable
- restricted end-use markets

The most sensible approach is to target specific streams where a better return on effort and finance invested can be expected (e.g. more focus on industrial packaging and easily extracted items such as detergent bottles).

In fact to go beyond the plastics material recycling levels cited in recent studies (around 15%) would require artificial funding mechanisms to attain the practical maximum. In addition alternative technologies and markets would need to be developed.

Two waste management routes remain unsupportable: landfilling, which is considered a true waste of the potential within plastics articles, and incineration without energy recovery, which again is suboptimal in terms of resource utilisation. Apart from these two disposal routes all other approaches have validity and the selection of the most appropriate method depends upon local conditions. Allowing the freedom of choice will enable more holistic solutions to waste management to be achieved.

This treatise, although not rigorous in scope and subject to qualified assumptions, has postulated that packaging waste would rise by almost 90% by weight in the absence of plastics, with a corresponding effect on energy utilisation.

Not only therefore are plastics indispensable in our modern lives, but because of their inherent low weight and high strength they provide enormous leverage as a packaging medium. Furthermore, returning to a pre-plastics packaging state is not actually an option because the degree of quality, hygiene, cost-performance and resource optimisation that is enjoyed today would be compromised.

In conclusion, plastics packaging will continue to contribute towards the optimisation of resources, helping to satisfy societal needs in an economic manner—in short, to contribute towards the goals of sustainable development.

Acknowledgements

The author extends thanks to D. Geden and D. Bain for support in preparing this chapter.

Bibliography

Packaging without plastics—ecological and economical consequences of a plastics-free packaging market, GVM (Gesellschaft für Verpackungsmarktforschung—society for packaging market research/D). Commissioned by VKE, the German Plastics Producers Association, Wiesbaden, 1992.

Ecoemballages, Ademe & Adelphe (1999) *Le gisement des emballages menagers en France*—donnees et references Via Reale, Paris.

M[c]Cormack, T. (1998) *Polypropylene in packaging*, seminar, RAPRA Technology Ltd, Shrewsbury.

The following six references may be found on the APME web site (http://www.apme.org), or by applying to Association of Plastics Manufacturers in Europe, Avenue E. Van Nieuwenhuyse 4 Box 3 B-1160, Brussels.

Plastics at work for a sustainable future, APME brochure, December 1999.

New insights into European waste management choices—assessing the eco-efficiency of plastics packaging waste recovery, APME Summary Report, January 2000.

Plastics—an analysis of plastics consumption and recovery in Western Europe 1998, APME Report, Spring 2000.

Plastics recycling in perspective, APME paper, May 1999.

Assessing the potential for post-use plastics waste recycling-predicting recovery in 2001 and 2006, APME Summary Report, June 1998.

Plastics a material of choice for packaging, APME brochure, April 1999.

10 Recycling and reuse of plastics packaging for the consumer market

I.S. Dent

10.1 The legislative pressures for recycling

From an environmental perspective, it has always been noticeable that the issue of design, manufacture and marketing of packaging (or more specifically packed products) is seen as an issue which is separate and far removed from the environmental considerations of the package when used. While it is understandable that the commercial pressures imposed upon packaging producers and users are of paramount importance for the viability of the industries that they serve, there has to be a growing recognition in all sectors of the packaging industry that the use of the packaging does not end with the customer. The imperative from the position of the environmentalists, both inside and outside government institutions, is that the responsibility for the disposal of used packaging, especially that from consumer markets, rests with those responsible for placing the packaging on the market in the first place. Thus, the revised marketing profile for the packaging industry is as shown in Figure 10.1. This figure originates from the German Environment Ministry in 1993 and is still relevant today.

Within Europe, there was the Liquid Beverages Directive in the early 1980s, and then the Packaging and Packaging Waste Directive (PPWD), which was finally adopted by the European Commission in December 1994. The former piece of legislation was focused directly on a specific sector of the consumer

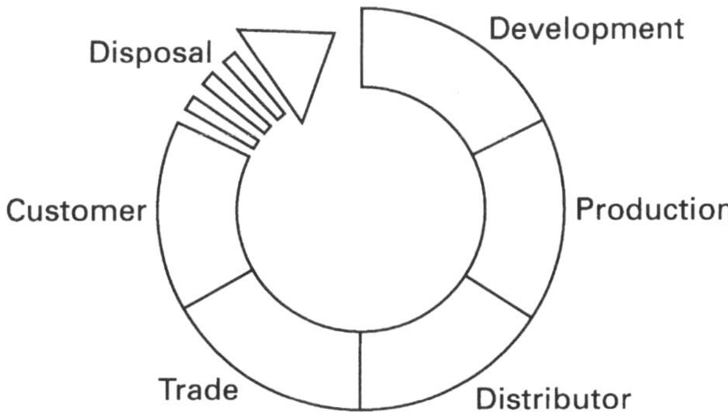

Figure 10.1 R & D/production/marketing/use/disposal.

market and although it failed as a result of its piecemeal implementation, it did show the intention of the politicians and the bureaucrats in Brussels in this respect. The challenge for users of plastics packaging to meet the PPWD targets should not be underestimated, especially in the UK (Figure 10.2).

Although there is much dispute over exact numbers, the issue is over the ability to manage the waste stream beyond minimum compliance and to establish a suitable recovery infrastructure beyond 2001, to maximise any cost efficiencies and to generate a successful and sustainable recovery and recycling culture.

The PPWD is the main driving force behind the recycling of packaging in Europe today, and this is likely to continue as the politicians debate the extension of this Directive after 2001. Irrespective of the various arguments for and against recycling, there is no doubt that the concept of 'producer responsibility' is becoming well established in Europe. The current debates in Europe on Integrated Product Policy (IPP) extend this concept, and will be the main issue facing producers.

Although the legislative pressures are a main environmental driving force, this should not overshadow the real economic forces which have driven the packaging industry, and especially the plastics packaging industry, to greater and greater minimisation of packaging in relation to the product carried. The exponential growth in the use of plastics packaging since the 1950s can be attributed to the potential to carry more product for less cost, thus more readily meeting the growing demands of packaged products since the start of the consumer boom. This has been coupled with the versatility of plastics packaging to the delight and advantage of the designer and marketer. All of these virtues, however, have contributed towards the current environmental pressures and shift in values.

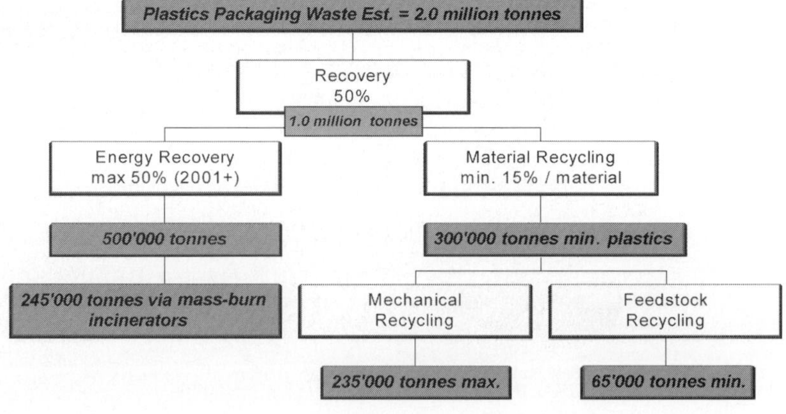

Figure 10.2 UK and PPWD targets to 2001.

It is important therefore that all engaged in the packaging industry understand *and listen* to the debate on packaging waste. Only by demonstrable activity in embracing these issues from design to marketing to using recycled material will the industry be considered as a suitable partner in a sustainable environment.

10.2 The sustainability pressures on packaging

Sustainability is the real key to the recycling and reuse of packaging. Packaging has become synonymous with a 'throw-away' society and thus not regarded as sustainable. Firstly, what do we understand by sustainable development? Although the term took on its first significant mantle in the Brundtland Report in 1987 [1], the broad definition of 'Meeting the needs of the present without comprising the needs of the future' did not have a major impact on industrial society until the late 1990s, when the Rio Conference started to address the need to translate the concept into concrete actions and definitions.

The third publication by Biffa on waste and resource management, *Great Britain plc—The Environmental Balance Sheet* [2], showed that consumption of energy and resources to manufacture products consumed in the UK was in the ratio of 10:1 (Figure 10.3). Of these products, approximately two-thirds are classified as consumer goods, including the 30 million tonnes for food; and nearly all of these goods require packaging.

The above calculation does not include use of water resources, which if included would increase the ratio to 100:1! Only by generating such statistics does the impact of sustainability take on a meaningful aspect. The recognition

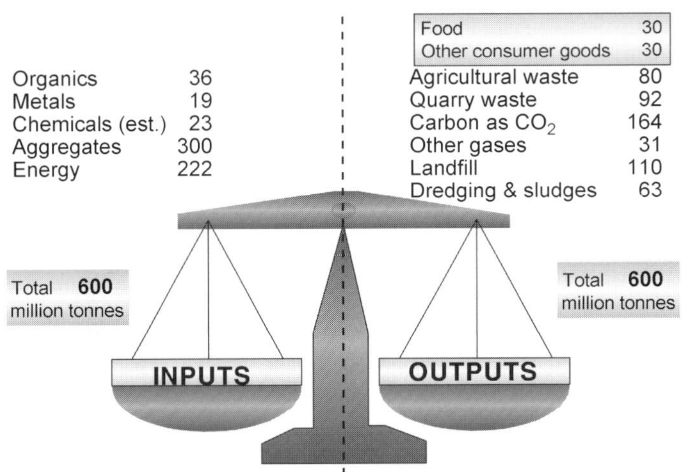

Figure 10.3 Inputs *vs* outputs of GB plc. Source: Biffa.

of examining this ratio of resource use to product generation was addressed by the Factor 10 Club [3]. They came to the conclusion that to be sustainable during a period when human populations are likely to double and average living standards increase significantly, industry needs to increase its resource conversion efficiency by a *minimum* Factor 4 (i.e. 75% reduction in resource consumption for any unit of production). Since industrialised western societies typically consume 20–30 times more than the less developed countries, they also concluded that here a Factor 10 (i.e. 90% reduction) was needed. Clearly, such radical changes will take decades to achieve and will have to be promoted by increases in the price of raw materials and tax reforms. For an overall appreciation of how business is responding to the sustainable development challenge at the end of the twentieth century, see *Cannibals With Forks* [4].

10.3 Plastics as a material of choice

Plastics as a material for packaging has distinct advantages over other materials in resource efficiency, since, irrespective of the previous arguments, its achievements in this sector have been attributable mainly to cost savings through its greater resource efficiency. Technological developments have created new generations of both commodity polymers like polyethylene and the more specialised polymers like polycarbonate. In all cases the drivers have been a need to compete in an increasingly competitive raw material market and to meet the demands of the user in seeking ever greater weight reductions, as demonstrated by Table 10.1.

Ironically, this trend works against recycling and reuse in many aspects: firstly, by encouraging the use of multilayer and multipolymer laminate structures to generate tougher but thinner substrates; and secondly overall lower weight in the waste stream.

Both factors mean that it becomes even more difficult to extract the waste from the household stream where the consumer packaging ends up and find suitable recycling technologies and end-markets. However, it is true to say that the plastics industry has been at the forefront of technological development in

Table 10.1 Resource efficiencies

Product pack	Reduction in weight 1970 to 2000 (%)
Yoghurt pots	60
Fizzy drink bottles	33
Plastic carrier bags	50

Source: INCPEN [5].

the mid to late twentieth century, and to achieve the Factor 4/10 results will require technological evolution above all else.

10.4 Life-cycle analysis

Increasingly, this dimension of resource efficiency and sustainability leads to the concept of life-cycle analysis (LCA) as a measurement tool to gauge the overall impact of a product, including its packaging and, more importantly, the improvements either past or potential.

Unfortunately, life-cycle analysis has not become a universally accepted measurement tool for several reasons. Firstly, the concept itself has been distorted by its application as a marketing tool against competition rather than as a resource efficiency tool. Secondly, it has taken many years for the practitioners to agree on the basic measurement criteria and parameters for energy, raw material and emissions data. Thirdly, the ability to generate data meaningful to the layman has been fraught with difficulty and has usually ended up generating more questions than answers. Until a more effective means of communication is found, LCAs will only be effective as an internal measurement tool for industry to seek ways of improving resource efficiency, usually based upon existing technology. Finally, and importantly for the subject of this chapter, the use of LCA as a measurement tool for focusing on reuse and recycling only will always be an imperfect platform, as end-of-pipeline solutions can only provide an answer from a particular viewpoint, not the whole spectrum. Thus, any reader wishing to understand and use life-cycle analysis should consult the Association of European Manufacturers in Europe (APME) [6], who have established pan-industry life-cycle inventory data (raw material usage and emissions) for all the main polymers, plus some conversion activities for film and bottle production. APME has also sponsored other LCA studies on plastics packaging waste, notably that based on the German plastics waste management system [7]. For more in-depth reading on the subject, especially in relation to the broader agenda of sustainability, two publications may be of interest. These are *Driving Eco-Innovation* [8] and *Who Needs It? Market Implications of Sustainable Lifestyles* [9].

10.5 Future legislation

The Packaging and Packaging Waste Directive has been further developed with the adoption of a ‘daughter’ directive, the Essential Requirements Directive. In the UK, this came into force in January 1999 and sets out the guidelines under which all packaging placed on the market must conform to specified requirements on reuse, recycling, recovery and composition. This is the essence of the environmental legislation on packaging— criteria designed to reduce the amount of packaging placed on the market and to ensure that packaging which is placed on the market meets required standards for manufacture and optimum recovery after use. No one designing, using, specifying and placing packaging on the market should ignore these regulations. They will be an effective tool in realising the vision of the German Environment Ministry, which is to ensure that basic environmental features in packaging are designed into the product, and that any packaging not meeting these requirements will be banned from the marketplace. The ability to reuse and recycle used packaging will be a key criterion as part of this suite of requirements, and is a fundamental challenge for the plastics packaging industry.

10.6 Plastics packaging recycling in the consumer goods sector

Recycling is not new to the plastics industry. Since the oil crisis of the early 1970s, all producers have been seeking to maximise the value of their waste streams. Thus, the production and industrial waste streams have seen some major growth in recycling activity, where the polymer composition, degree of contamination and logistics make the exercise economically viable. By the nature of these factors, it is not difficult to see that the predominance of the plastics recycling activity is in the non-consumer areas (Figure 10.4). Finding solutions which are economically as well as environmentally sound poses difficult problems for an industry which is cost-driven and used to seeking cost reductions, especially for packaging, which is only seen as a functional unit in the total supply chain of a product.

Developments in the recycling of consumer packaging have been largely confined to the household bottle sector, where the weight, quantity and limited polymer range allow for a more convenient recovery system. However, this has been piecemeal, depending on the legislation in place and the support costs from the packaging industry— either voluntary or mandatory. While the environmental theory and environmental balance of such activities are usually positive, the economics are usually less positive. End-markets for the recovered material vary according to the price of the virgin polymer, to the extent that during periods of price troughs it is more economical to use virgin material than recycled material (Figure 10.5). There is a fixed cost to mechanical recycling below which it is uneconomical to sell the recyclate and the process is no longer viable.

Like all material recycling, the problems are often caused by the fact that waste generation is local by nature, while the recycling activity is more regional or even national, and the end-markets may be affected by global factors. This is especially the case for PET (polyethylene terephthalate) bottles, which are mostly turned into a flaked form and then into a fibre end-market. Owing to

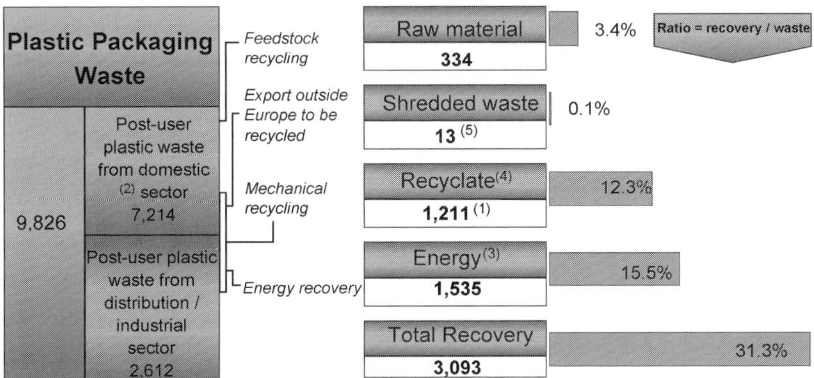

Figure 10.4 Total recovery activity (Western Europe, 1997).
Source: APME SOFRES
(1) of which 208,000 tonnes from intra Western Europe trade. Imports from outside Europe are not taken into account
(2) MSW/household
(3) The figure was re-evaluated taking into account the real quantities incinerated compared to available capacities of incinerators
(4) Waste supplied to recyclers
(5) Western Europe exports 13'000 tonnes as a whole.

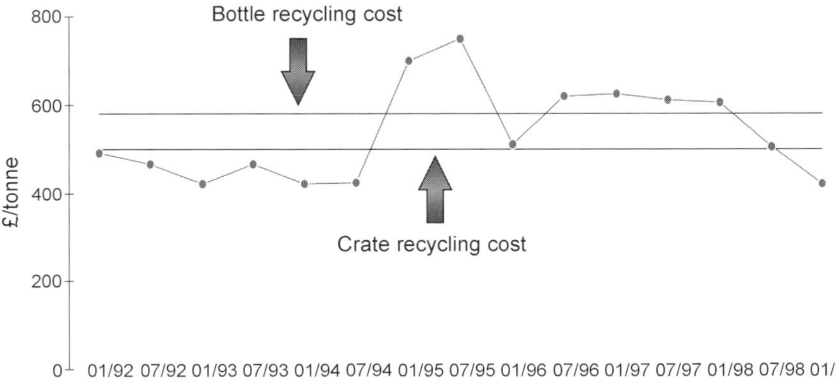

Figure 10.5 Recycled HDPE price trends. Source: Linpac.

the requirement for high purity polymer flake, the recycling of the bottles is a demanding reprocessing stage and thus carried out either by the fibre producer or selected recycling facilities. Use in the fibre market is also affected by global activities on fibre production, where competition is intense between the Far East, the US and Europe. The price for PET bottle polymer is, however, higher than that for PET fibre, and thus can withstand the economics of recycling much better than the other polymers, where price differentials do not exist to the same extent.

The other effect, becoming more noticeable in the drive to meet the legislative targets in Europe, is that the export markets provide an easy solution, especially those in the less-developed economies where lower specification products are more acceptable. The overall impact of this has still to be fully felt, but already there has been a backlash against Germany exporting its used plastics packaging to neighbouring countries like France and distant markets like Indonesia and China. Germany received much attention as a result of its early adoption of packaging waste legislation in the early 1990s, but as more EU Member States implement the PPWD, exporting has increased. This is not just in consumer packaging, but also in the commercial packaging sector, especially the film stretch/shrink-wrap around pallets (Table 10.2).

The inevitable consequence of the above is that as all EU countries try to reach the present minimum 15% recycling level, the export market will be saturated and thus countries will have to compete more aggressively on price or seek solutions nearer to home. The latter solution will not be cheap as this will involve investment in the infrastructure and capital expenditure. Germany has already had to face this situation and those countries which plan for this eventuality will be better placed both to meet the targets and to provide more economical solutions in the long-term.

10.7 Recycling in the US and Japan

Recycling does not receive the same kind of profile outside Europe. In the US, the activities and initiatives to promote recycling are conducted at state

Table 10.2 Import/export activity of plastics packaging waste: European plastics recycling rates 1996

	Local recycling		Recycle rates		
Country	Domestic ×1000 tonnes	C & I ×1000 tonnes	0 %	1 %	2 %
Germany(*)	433	133	37.1	37.1	45.4
Belgium	1	21	6.5	30.4	11.8
Netherlands	10	48	11.1	19.9	12.8
France	18	102	7.4	8.3	9.6
UK	8	79	5.9	6.5	6.1

Note: *Feedstock recycling include = 251'000 tonnes.
Recycle rates: 0 = Local recycling volumes only
1 = Apparent recycling of country (including imports)
2 = Apparent ability to manage waste (no exports/imports)
Source: APME.

level, with states such as Oregon and California being more demanding in their requirements. These states use mandatory recycled content legislation as a key force. This legislation is also heavily focused upon consumer packaging, but has to recognise the fact that use of recycled material in food-contact applications is severely limited depending upon the recycling technology applied, and thus limited in its success on a wide scale.

The US does benefit from its application of waste management practices to the total waste stream rather than to specific market sectors as in Europe. Thus, the recovery and recycling of plastics itself is the main criterion, not whether the piece of plastics has been used for packaging or not. This allows for much better economies of scale in the logistics, which is where much of the costs are generated with lightweight materials like plastics. The only real activity on plastics consumer packaging in the US is in the PET and HDPE bottle markets, where economics are recognised as being more favourable.

The Japanese also appear to adopt a much more pragmatic approach to their waste streams. The attitude towards the recovery route for energy utilisation is also more relaxed and prevalent. The Packaging Container Law passed in the late 1990s was Japan's official recognition of the trends in Europe. However, in the case of plastics the Government recognised the difficulties and settled for a declaration on recycling targets for PET containers only, with the rest of the targets focused generically on the use of feedstock recycling technologies for conversion and use of mixed plastics as a raw material in upstream processes. They have concentrated their effort on the use of plastics as a hydrocarbon base for blast furnaces—a subject which is covered along with other feedstock recycling technologies later in this chapter.

10.8 Plastics packaging reuse systems

One of the most striking developments in the past decade in packaging is surely the change in the washing detergent market. It has progressed from a cartonboard pack containing a powder, to a plastic rigid container containing a liquid, to a plastic refill pouch, to a liquid concentrate in a composite pack, to solid tablet form in a cartonboard container with a nylon mesh holder. All of these changes have been driven by economics but in the process have reduced the amount of packaging on the market in the same period (Table 10.3) and reduced the consumption of resources, since the use of concentrates and solid tablets means that less water is transported.

In the scale of environmental improvements, reducing use of resources at the production stage as indicated earlier has a more beneficial impact than the recycling of a greater amount of packaging. However, it is still necessary to ensure that the package is capable of recovery or recycling at the end of its

Table 10.3 Packaging reduction in the detergent market: reduction of packaging of concentrated detergents 1991 to 1994

Year	Packaging × 1000 tonnes	'Virgin' packaging material × 1000 tonnes
1991	80	20
1994	67	13

Source: Lever Europe.

life—and therein lies a dichotomy, since quite often to reduce the amount of packaging as in the refill pouch example a barrier has to be used to contain the liquid adequately. This in turn renders the package non-recyclable, at least by mechanical means.

It is worth repeating therefore that in order to appreciate the value to the environment of any packaging change, a full life-cycle approach is required. The major producers like Proctor and Gamble and Lever do undertake such LCAs on their product pack changes to assess the economic and environmental impact. Such analyses will become commonplace, with more user-friendly software programs being available to producers and specifiers to appreciate their impact.

A minor revolution in refill development is taking place in the household chemical cleaning market, where spray pumps can be interchanged in refill containers. There is much more scope for such innovations in improvements in the smaller plastic rigid container market.

The most popular reuse application of course is the ubiquitous plastic carrier bag with the promotional opportunities created by retailers such as Sainsbury's for money-back schemes. Even here of course, the plastic carrier bag is giving ground to the sale and reuse of plastic rigid boxes for shopping. Interestingly, there does not appear to be a very visible take-up of this new system, perhaps because it takes more effort and cost for consumers to use boxes instead of carrier bags. To gain consumer support a system needs to be convenient. The plastic carrier bag not only has the appeal of being free, but if necessary can be reused in the home for waste-bin liners or for carrying other items. The rigid boxes are specific in their use, cost money and require a dedicated routine to benefit from their purchase.

10.9 Consumers and reusables

The attitude of consumers is of paramount importance for packaging, especially for plastics with their lighter weight and diversity of application. This attitude can also extend to the refill market, where consumers find it more convenient to use a new pack than put themselves to any inconvenience by refilling a container. This was the case with the plastic refill pouches for liquids. The only real impact on consumer attitudes will be cost; that is, if there is a greater differential between a new pack and a refill to warrant the additional effort. Because of the efficiency of the packaging systems and their overall cost in the product value chain, the differential will not be so great—unless an environmental tax or levy dictates such a differential. This may be an option facing producers in the future.

Refillables are becoming of greater predominance in countries such as Germany and Austria, where quotas are now imposed. However, like all systems, the packaging is only one element and thus when imposing any quotas by governments, due recognition needs to be given to the whole system, not just the pack. Different schools of thought exist on the overall benefits of refillables compared to one-way packages, but this can only be decided on a case-by-case basis. While refillables can generate innovation, as demonstrated by the detergent market, it can also stifle innovation if that is the only criterion allowed.

Much has also been made of refillable systems such as the deposit and return systems employed for liquid beverages in the past. However, times have changed in both the sophistication of the packaging systems and the shopping habits of the public. Despite the advent of loyalty cards in most stores, the flexibility of when and where to shop will determine the majority of consumers' attitude towards deposit and return systems. The Body Shop has been quoted as stating that the greatest take-up of their introduction of a refillable system was 12% at any one store, which given the profile of their usual customer must seriously question the use of such schemes in today's environment. One development which may alter attitudes is the advent of electronic shopping. If it develops as planned with deliveries being made to the householder on a regular repeat basis, then a deposit and return system may be more effective—if the retailers can develop a suitable infrastructure. One key issue concerns contamination of used packaging in the same vehicle space as new food and products. However, this problem has been overcome by the Save-A-Cup scheme for recycling of used plastic vending cups with the introduction of shredders to remove the contamination and render the plastic sufficiently clean to be carried with virgin vending cups. Could we see the advent of shredders in the normal household for such applications in the future? It would certainly reduce the volume in the waste bin and provide a ready means of return. As usual there is always a downside, as this system could only work if technology allows for the ready separation of the shredded material into the various polymer fractions (see the APME study on sorting techniques [10]) or recycling of mixed plastics becomes sufficiently well-established to make the operation viable. The range of options on recycling mixed plastics is covered later in this chapter.

10.10 Plastics recycling technology

All recycling irrespective of material type requires three main ingredients to become economically and environmentally sustainable:

1. sufficient quantity and quality of material as input to the process
2. ready end-markets and
3. sufficient economic drivers from waste collection to resale.

In the case of plastics recycling, these conditions have until now been met for some commercial and industrial recycling (i.e. some film fractions and crates). The main difficulty lies in the first element of being able to generate sufficient quantity and quality of a homogeneous polymer, which is usually required for most current mechanical recycling operations. The situation is considerably aggravated when considering plastics packaging in the consumer market.

Figure 10.6 shows that approximately two-thirds of the applications for plastics packaging in Europe lie in the domestic sector. Of this, approximately 25% is accounted for by the household bottle market in the 1–5 l range, where the average weight is 20–40 g. The average weight of other items of plastic packaging in the consumer market is in the region of only 5 g. If the first of the criteria listed above is taken for successful recycling and applied to Figure 10.6, it can be calculated that only 25% of the total is suitable for collection and recycling as a homogeneous and non-contaminated source.

Thus, the ability to meet the minimum recycling target by the traditional mechanical recycling routes is unlikely to be sufficient. There is wide variation in the various EU countries, but the APME have recently conducted an extensive study into this aspect (Table 10.4).

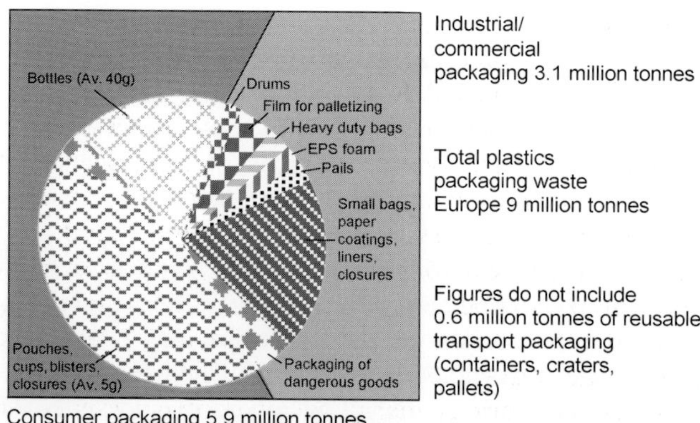

Figure 10.6 Plastic packaging applications in Europe.

Table 10.4 Potential for mechanical recycling Western Europe (1995, 2001, 2006)

	Total waste	Waste collectable for mechanical recycling	Recycled resin demand	Potential for mechanical recycling		
	tonnes ×1000	% of total waste	% of total waste	tonnes ×1000	% of total waste	AAGR in %
1995	16 056	7.6	–	1 222	7.6	+9.5
2001	21 288	11.3	12.9	2 105	9.9	+9.5
2006	25 751	12.7	12.9	3 723	10.6	+5.3

Source: APME [11].

The main conclusion of this report echoes those sentiments mentioned earlier on exports: 'Mechanical recycling of post-use plastics packaging has the potential to increase to a European average of 14.8% (1.8 million tonnes) in 2001.... However, there is an upper limit to potential demand for recycled plastics. Countries achieving high mechanical recycling levels can export recyclate to neighbouring countries. Once these importing countries begin processing and selling their own recyclate in greater quantities, then saturation may be quickly achieved.'

10.11 Technical problems in mechanical recycling

As readers will probably appreciate, being involved with plastics packaging, the great benefit of plastics as a packaging medium is the ability to create many different specifications from a variety of basic polymers to suit the production parameters required. The difficulty lies in being able to utilise these polymers in a heterogeneous mix when discarded, especially if residues remain as additional contaminants. The use in a physical form, as in mechanical recycling, means that the basic physical properties are still inherent in the polymers. The presence of polymers with different physical properties means that during processing the polymers will not readily mix and thus form a polymer compound of guaranteed physical and mechanical properties. The specifications even for PET, PVC and HPDE bottle recyclers, especially PET, is testament to this problem. Without the guarantee of quality, the recycled material is not suitable for resale. The sale of recycled material in whatever form and by whatever route needs to meet certain physical specifications, even if not as stringent as virgin polymer. Consistency is the key, which is unlikely to be met if the raw material sourced is of a constantly changing mixed polymer origin, as is usually the case with a mixed plastics stream from domestic waste.

There are of course various means for overcoming these problems in a mechanical operation, for example, the use of compatibilisers. Much effort has been expended on overcoming the problem of mechanical recycling of heterogeneous polymers, some with more success than others. The use of compatibilisers is often a problem because of cost rather than technical performance as the compatibilisers themselves are expensive. Different production techniques have been applied for mixed plastics, like compression moulding, usually with a synthetic lumber application as the end-market. However, the markets for synthetic lumber are limited and quite often while the product is superior for durability, especially in external applications, the overall costs compared to traditional wood make the product uneconomic. The lower technical specification end of the market once again has limited application, especially in Europe. There are markets for these products overseas, where standards in production processes and use can vary considerably from those in Europe.

10.12 Feedstock recycling

The scale of the problem is highlighted in the APME report [11]. The study does indicate that if Europe is to tackle the immense scale of the waste management issue, it needs to address the use of feedstock recycling technology more vigorously. Feedstock recycling or raw material recycling is the conversion of plastic waste back to its intermediate or primary building blocks through chemical or physical processes. Feedstock recycling differs from mechanical recycling mainly in that the physical properties of the waste polymers are not retained, and recovered chemicals may be purified and made into new products, with properties identical to virgin products. Hence, the end-market and reuse problem is removed. If the process can tolerate a mixed plastics waste stream, one of the other three criteria for successful material recycling is also removed. This leaves the third criterion— economic drivers. So far most processes have been pioneering and thus costly in research and development terms. The only successful applications have been those employed in Germany, where the economic and environmental drivers were in place to tackle the capital and operational support required on a commercial scale. These were built upon existing facilities, which could be adapted to use a mixed plastics stream as a raw material source.

Different recycling techniques are now examined in more detail. Figure 10.7 shows in a simple manner the basic differences between the two types of plastics recycling. This does *not* mean that they are competing however—far from it. Where the environmental balance favours mechanical recycling, as is often the case with bottle recycling and clean polyethylene film, then it should be used and encouraged.

10.13 Mechanical recycling technologies

In the case of commercial/industrial recycling where a single source of polymer can usually be guaranteed, including any product contamination, the sorting

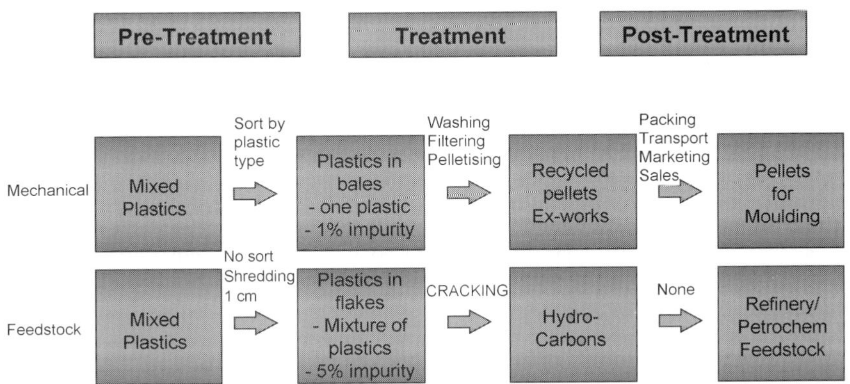

Figure 10.7 Mechanical against feedstock recycling.

stage is minimal. Washing and drying is not always carried out, except in the case of post-use film reprocessing. Unlike most other production processes, recycling cannot rely upon quality certification for its raw material input; therefore a certain element of the production process involves sorting to ensure the required quality. This is particularly true for consumer packaging even if the material has ostensibly been pre-sorted. For instance, the LCA study carried out in Germany on the DSD system for plastics packaging [7], where the material is all from consumer sources, showed that the quality was as shown in Table 10.5. Obviously the bottle/container fraction should be the easiest visible source, but the above figures do not include the additional sorting required into PET, PP, PVC and HDPE fractions for mechanical recycling.

The impurities of different polymer fractions can be sorted by the use of cyclones and flotation processes in the washing and drying stages, which of course is an extra cost to that usually experienced with industrial and commercial recycling operations.

Once the product has been suitably purified and dried, it is then processed into a suitable form for resale and use. This may vary according to the polymer and application; for example, in the case of PET for fibre production, the flaked form is sufficient, while in the case of HDPE the pelletised form is usually

Table 10.5 Quality of household plastics waste

	Sorting yield (%)	Purity of sorted fraction (%)
Film	9	95.2
Bottles/containers	7.5	99.0
Mixed plastics	58.4	88.6

required. This means that after washing and drying the polymer is shredded and reprocessed in an extruder. Thus, additional costs are incurred.

This cost differential (see Figure 10.8 and Table 10.6) was also highlighted in a paper presented by VKE, The German Association of Plastics Manufacturers in 1997 [12].

When looking at the mechanical recycling of plastics packaging, there is a need to understand the different fractions and the applications in order to appreciate the costs structures involved. This does not mean that the problem

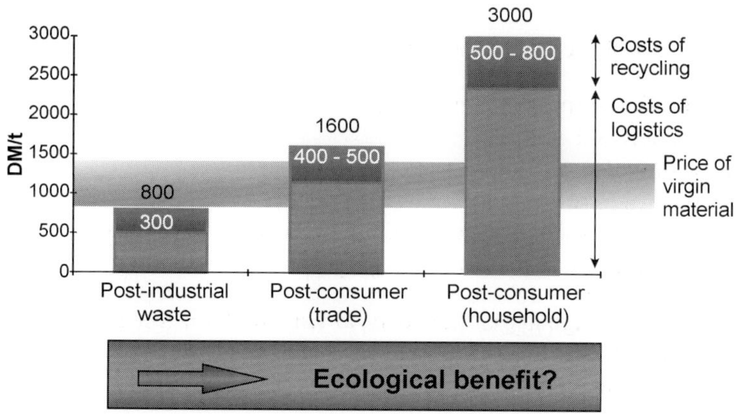

Figure 10.8 Dependence of waste management costs on waste quality plastics packaging. Source: VKE [12].

Table 10.6 The economics of plastics recycling

1. Costs (per tonne)		
	Post-consumer e.g. bottle (£)	Post-commercial e.g. crate (£)
Collection/sorting	175	125
Transportation	20	50
Yield loss	40	10
Recycling	360	265
Total	595	450
2. Typical recyclate prices (April 1998) (£)		
HDPE	425	
PP	375	
PS	450	
3. PRN price		
Reflects costs—REVENUE		

Source: Linpac.

should be ignored or used as an excuse for inaction—it is a challenge to seek more cost-effective routes and alternative methods.

The methods for reprocessing in mechanical recycling applications and the problems encountered in extruding and injection moulding recyclates, plus those associated with homogeneous and heterogeneous streams, are covered in detail in a publication by the German Plastics Manufacturers and the German Institute for Plastics Processing [13].

10.14 Feedstock recycling technologies

The advantages of feedstock recycling are:

1. the ability to manage a mixed plastics waste stream
2. unlimited end-markets
3. a variety of potential technologies
4. large-scale production requirements and
5. the possible provision of a universal solution to all waste plastics management, not just that of packaging

The main disadvantage is that the costs, at least in the initial stages of research and development, are high. It does, however, offer the plastics industry the opportunity to manage its waste stream in a similar manner to that of the other material streams.

The technologies developed so far can be categorised into solvolysis and thermolysis routes.

Solvolysis. The critical factors are:

1. suitable for condensation polymers (acid + glycol = polymer + water), e.g. PET, polyurethane (PUR), polycarbonate (PC), nylon
2. requires pure polymer stream
3. chemical reactions
4. produces higher value products and
5. high preparation costs

Thermolysis. The critical factors are:

6. suitable for addition polymers, e.g. polyethylene (PE), polypropylene (PP), polystyrene (PS), PVC
7. allows mixed (dirty) stream
8. thermal treatment employed
9. lower preparation costs and
10. resultant products lower value than solvolytical techniques

As a result of the predominance of the polyolefins (PE, PP, etc.) in packaging—approximately 75% of total polymers—the use of thermolytical techniques has so far received the most attention.

10.14.1 Solvolysis routes

Techniques in this category include (Table 10.7):

- PET— employing glycolysis, hydrolysis, methanolysis;
- polyamides— employing hydrolysis;
- polyurethanes— employing hydrolysis and alcoholysis.

Only the PET sector is examined in detail here. However, for more detail on the other technologies, readers are recommended to consult *Recycling and Recovery of Plastics* [13]. The chemistry involved in the solvolytical routes is shown in Scheme 10.1

Table 10.7 Solvolysis routes

Route	Feedstock quality	Intermediate product	Capital	Conversion cost	Comments
Methanolysis	Low	DMT	V. high	V. high	Capital intensive
Glycolysis	High	Oligomer	High	V. high	\$48–72/kg premium over virgin
Partial glycolysis	Low	PTA	V. high	Highest	No commercial facility in world

Source: *Packaging Week*, November 1995.

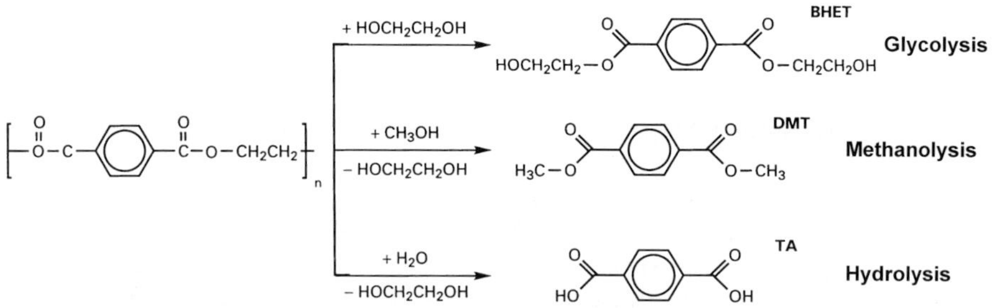

Scheme 10.1 Solvolysis of PET. BHET = bis(hydroxyethyl) terephthalate; DMT = dimethyl terephthalate; TA = terephthalic acid. Source: *Recycling and Recovery of Plastics* [13].

10.14.2 Thermolysis routes

The scale of research in this area is shown in Figure 10.9 and Table 10.8.

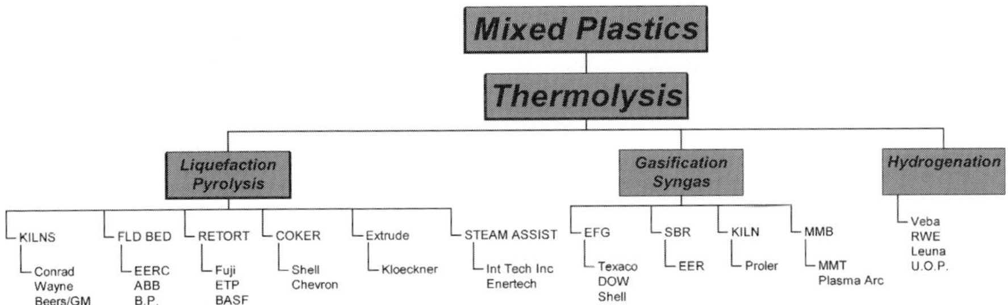

Figure 10.9 Feedstock recycling of mixed plastics. Source: *Feedstock recycling of mixed plastics* [14].

Table 10.8 Global research underway in 1996

1. Europe	
Kilns/retorts	Veba, BASF
Fluidised bed	BP & Consortium
Degradative extrusion	Kloeckner, Leuna-Werke
Hydrogenation	Veba, RWE
Gasification	Veba, Shell
2. US	
Kilns/retorts	Conrad, Wayne, GM
Fluidised bed	EERC, Battelle
Gasification	Texaco, EER, MMT
3. Canada	
Kilns/retorts	Vacuum pyrolysis
Hydrogenation	CanMet
4. Japan	
Kilns/retorts	Fuji Tech

This is not an exhaustive list, as many universities have also conducted research in this field and it may continue to attract research funding, but the position in 1999 has seen a reduction in the number of organisations involved. In Europe, this is due to the development and application of the technology on a commercial scale (with subsidies) in Germany, with only the BP process and Texaco process from the US being serious contenders outside Germany.

In the US, the different (i.e. lower) legislative pressures and geographical spread of petrochemical plants and refineries have reduced the level of research. The preference for mechanical recycling of bottles only in the main active States has also reduced interest in developing research further.

In Japan, the emphasis is still very much upon feedstock recycling as indicated earlier for all mixed plastics excluding PET. However, the main emphasis is on the reductant route in iron making, for which a licence has been taken out from

DSD, Germany, to utilise the same agglomeration preparation route and feed for NKK, the Japanese steel-maker.

10.15 The German experience

Since Germany introduced environmental legislation earlier than other European countries, generating substantial funds in the process, this is where most experience of plastics waste recycling is found. The techniques employed, however, are those built on existing production units, and thus not easily transferable. The only transferable technology is that using plastics waste as a reducing agent in iron making.

Thermolysis routes have been used so far, since the major components of the waste polymer streams in packaging are polyolefinic in nature and hence addition polymers.

The German experience has demonstrated how the use of a mixed plastics waste stream does require some preparation to feed the product into the various industrial processes adequately. It should be remembered that none of these processes are akin to energy recovery routes, especially mass-burn incinerators where waste can be accepted into the feed operation in a crude form. Like most industrial processes, there are specifications for raw material input developed over many years to suit particular operations; in the case of feedstock recycling, where the waste plastics is largely replacing an existing raw material at varying rates, then the specification of the waste plastics must meet both the physical and chemical properties of the existing raw material. In Germany, the feedstock recycling processes are mainly prepared to an agglomerated form. This is all managed by DSD, who contract various sorting centres to take local authority waste and sort and bale the plastics material to a required specification (i.e. no non-plastics material). The material is then delivered to preparation plants that are once again contracted to DSD. After checking the quality of the bales at these preparation plants, the material is then fed into a large trommel machine for coarse separation and then eventually through an agglomerator (compactor), where mild heat mainly from the process binds the material together. This is necessary as the film fraction of the waste plastic does not lend itself to simple shredding or flaking for feeding into the feedstock recycling processes, as would the rigid plastics fractions alone. The volume being utilised in this manner in 1997 was:

- Hydrogenation: Veba 70 000 tonnes per annum (tpa)
- Gasification: RWE 100 000 tpa

There is also a major development at Stahlwerke Bremen, where the plastics waste is used as a reductant in the blast furnace.

10.16 Use in blast furnaces

This deserves a special paragraph, as it is using plastics waste as a hydrocarbon source, but not in an oil/petrochemical upstream process as with other technologies. Because of this, there is some debate outside Germany on the definition of this route in terms of recycling or recovery. However, in simple terms, while recovery operations utilise the hydrocarbons from the plastics to release them only as heat or power, the blast furnace *requires* heat to generate the process and uses the hydrocarbons in a material form as shown in the chemical equation below.
Injection of waste plastics (+ oxygen):

$$C_2H_4 + O_2 = \underset{\text{(syngas)}}{2CO + 2H_2}$$

Reduction of iron ore to pig-iron:

$$4Fe_2O_3 + 8CO + 4H_2 = 8Fe + 8CO_2 + 4H_2O$$

The actual process is shown in Figure 10.10.

As a precursor to steel production, the raw material, iron ore, is converted to pig-iron. As the iron ore contains chemically-bound oxygen, this needs to be released by the reduction process shown above. The hydrocarbons, either in the form of heavy oil, pulverised coal or waste plastic, are injected at the *bottom*

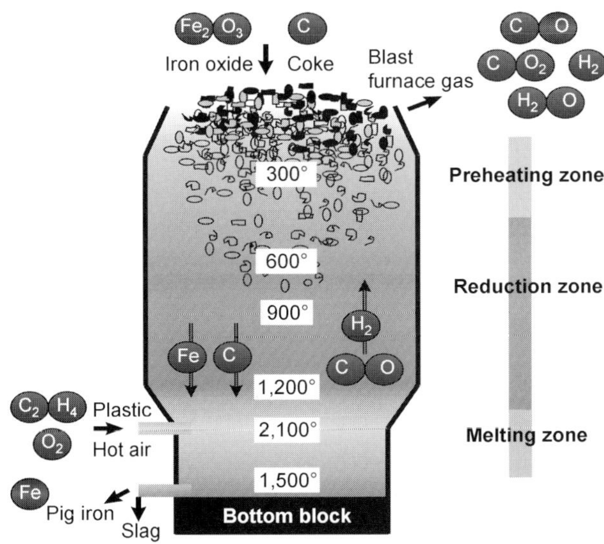

Figure 10.10 Blast furnace. Source: DSD, Germany.

of the blast furnace to complete the process of reduction, which has already commenced with the feed of iron ore and coke at the *top* of the furnace. The hydrocarbons injected in the raceway at the bottom of the furnace at 2000°C immediately form syngas, a mixture of hydrogen and carbon monoxide, which in turn reacts with the iron oxide to remove the oxygen. As the gas migrates up through the blast furnace shaft, more than 80% of the reduction potential of the gas is used up, with the remainder being released at the top of the furnace as a mixture of carbon dioxide, carbon monoxide and steam. Once the water has been removed the gas is used in the steel-works.

This process formed the predominant route for the disposal of mixed plastics waste in Germany by the end of 1998 (i.e. approximately 100 000 tpa). Extensive tests were carried out both prior to and during the operation of this process at Stahlwerke Bremen to ascertain the correct handling and feeding requirements and any potentially negative environmental impact (e.g. dioxins and furans). No adverse effects have yet been detected in this area. There is a limit on the amount of chlorine in the feedstock because of the corrosive effects of hydrochloric acid on the infrastructure of the furnace.

10.17 Other feedstock recycling developments

As indicated above, outside Germany, the two main developmental activities being pursued are the polymer cracking process initially developed by BP and the Texaco process, now centred upon their operation in the Netherlands.

10.18 Polymer cracking process

The polymer cracking process was developed initially by BP Chemicals in the early 1990s. It is based on unzipping the polymer chains back upstream to the petrochemical refinery stage. It is the only technology seen as a pan-industry solution and hence transferable technology on a wide scale. The technology became part of a consortium development with the first phase being financially sponsored by BP, Elf-Atochem, DSM, Enichem and Fina. The Fina portion was subsequently superseded by the APME, an organisation representing all the major polymer producers in Europe. This alone differentiates the technology from all the other feedstock recycling technologies.

The basic process is a simple low temperature pyrolitic technique using fluidised-bed technology, as shown in Figure 10.11.

A pilot plant to prove the technology was commissioned in late 1994 at BP Chemicals petrochemical complex in Grangemouth, UK, and successfully concluded in mid-1998.

After several runs of post-use mixed plastics it was demonstrated that the technology was capable of producing a suitable product for co-feeding into a

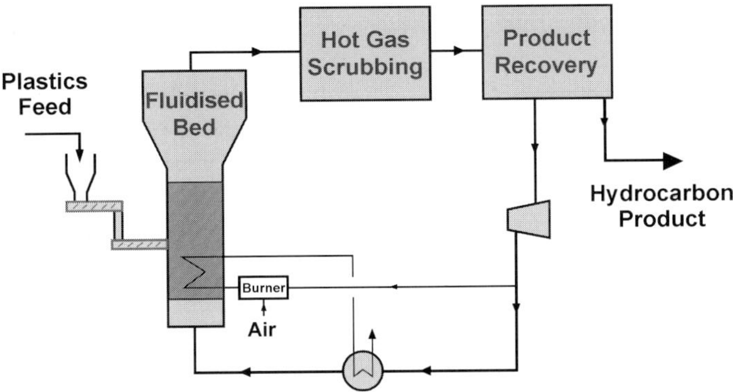

Figure 10.11 The polymer cracking process. Source: BP Amoco Chemicals [14].

refinery [15]. Since that period, the commercialisation of the process has been explored, notably with the support of Shanks (formerly Shanks & McEwan) and Valpak, the UK's largest compliance scheme on packaging obligations. BP Chemicals pilot plant is based in Scotland as is Shanks' northern operation, conveniently bringing together in one location the technology provider and the waste management company. The first phase of the commercialisation study is to examine the logistics, including preparation, to establish the feasibility of generating consistently a minimum 25 000 tpa mixed plastics waste and the elements required to produce the necessary quality of feedstock.

The technology is applicable throughout Europe and benefits from the fact that it is designed to be smaller scale than the other feedstock recycling processes (i.e. typically 25 000 tpa), which allows for more regional sustainability, given the relationship of most petrochemical complexes to the main conurbations in Europe.

However, the scale of economics also has a part to play in sustainability, and thus the eventual scale of operation needs to match the economic viability. This became apparent in Germany when BASF attempted to establish their technology commercially. While the technology itself was robust, the scale of the operation economically demanded a minimum of 150 000 tpa of waste plastics, which could not be guaranteed by the DSD.

The balance between economic and ecological factors needs to be addressed, and may be different for each EU country, but the plastics industry can find solutions which match the different criteria from the various alternative technologies now on offer.

Given the current non-transferability of the other technologies, except for the blast furnace route, in Germany, the plastics industry does need alternative solutions, and the polymer cracking route may provide one in the near future. There is a great need in the UK, for example, for alternatives to mechanical recycling.

10.19 Texaco gasification process

As referred to earlier, the Texaco development based on a gasification process was initially researched in the US, but interest has now switched to their European plant in the Netherlands. This process is more akin to those in Germany; that is, not readily transferable and based upon a commercial opportunity to utilise a mixed plastics waste stream as an alternative feed, rather than the polymer cracking process specifically tailored to a pan-industry use of plastics waste.

The gasification process is basically the production of synthetic gas (syngas, a mixture of carbon monoxide and hydrogen) as utilised in the blast furnace process. In a petrochemical process, the syngas can be used as a feedstock for the production of methanol, oxo-alcohol or hydrogen, or it can simply be burnt to power gas turbines for electricity production. The range of options has led the European Commission to debate the use of this technology for feedstock recycling accreditation.

Although the Texaco process in the Netherlands is linked to an ammonia synthesis of methanol, the main solid product is a vitrified slag which can be used in building products. The initial concept of the Texaco operation was a joint venture with Air Products, Roteb and VAM—the so-called PAX Rotterdam project. Air Products and Roteb were to be beneficiaries of the output of the process, while VAM, as the waste management partner, was to be the main provider of the input. However, the VAM link in the chain collapsed on economic grounds, despite this process offering ostensibly attractive economics. Hence, Texaco are now seeking alternative waste management partners in Europe. Only Finland has so far responded favourably. The scale of economics demands that the plant seeks a minimum requirement of 40 000 tpa, although more ideally 75 000 tpa, of plastic waste.

10.20 Degradability

This chapter on reuse and recycling would not be complete without a reference to degradability since the official European Directive on PPWD includes organic recycling— or composting—in its definition of material recycling. The issue of degradability in plastics is often misunderstood on both emotional and technical grounds. There has been much misinformation since the 1980s concerning the use of this technology in solving the plastics waste management problem.

The concept of plastic being degradable (i.e. disappearing on its own in a short space of time) is based on good intentions but poor technical understanding of the biodegradation process. The value to the industry of plastic is its ability to be durable and weather-resistant. For this reason, additives such as antioxidants and stabilisers are intentionally incorporated. Thus, for a polymer to become degradable means foregoing these attributes. Furthermore, in the waste management plan, the ability of plastics waste not to degrade or to degrade very slowly, is a *benefit* to landfill operations, where stability in the mass composition avoids subsidence and the release of gases and liquids to the environment. To be truly effective, degradability needs to occur in a composting environment, not a landfill environment. In fact, landfills are designed to preclude some of the main ingredients for degradability, such as light and oxygen.

From a technical standpoint, the ability of plastics to degrade needs to be restricted to those applications where a short life can be guaranteed or beneficial, and disposal is in a true composting environment. Both of these parameters pose significant hurdles. Firstly, how does a producer guarantee that the product is to be used solely for short-term applications, given the versatility of plastics use? Secondly, how can a producer guarantee that the product is to be used as intended? For example, plastic carrier bags can be used several times beyond the first shopping use. This is not to say that there are no niche markets for degradable polymers. Medical applications, such as sutures, are an excellent example. However, degradable polymers are likely to remain niche market products rather than a solution to the general plastics waste management issue.

Technically, a key issue is over what can be regarded as degradable. The incorporation of starch into polymers became a favourite topic when Italy decided that non-degradable carrier bags carried a levy *vs* degradable carrier bags. Almost overnight, the incorporation of starch into polymers became fashionable to allow the use of degradability on the carrier bag. A carrier bag retained for seven years lost its colour but not its mechanical strength despite the declaration, ‘This carrier bag is photodegradable’. Any starch if incorporated would degrade as would the pigment, but the long-chain polymers would remain or undergo slow degradation to short chains.

The main problem with the production of degradable polymers is the economy of scale, since the applications are generally for niche markets, despite the overt claims. The volumes are relatively small compared to the total packaging market, and production of genuine degradable polymers based upon polylactic acid and other renewable materials has not been developed as a large-scale operation, therefore the prices of degradable polymers are often three to four times higher than those of conventional polymers. Nevertheless, degradability is a popular concept and will continue to attract attention. The often exaggerated claims of the market analysts will promote this attention.

Bibliography

1. World Commission on Environment and Development (1987) *Our Common Future*, Oxford University Press, Oxford.
2. *Great Britain plc—The Environmental Balance Sheet* (1997) Biffa Waste Services of Severn Trent pk., High Wycomb.

3. Factor 10 Club (1995) *Carnoules Declaration*, Wuppertal Institute, Germany.
4. Elkington, J. (1997) *Cannibals with Forks*, Capstone Publishing Ltd, Oxford.
5. INCPEN Pack Facts (Minimised), Industry Council for Packaging & the Environment, London.
6. Eco-Profile Reports, APME, Brussels.
7. Arbeitsgemeinschaft Kunststoffverwertung (1997) *Life-Cycle Analysis of Recycling and Recovery of Households Plastics Waste Packaging Materials*, APME, Brussels.
8. Fussler, C. and James, P. (1996) Driving Eco-Innovation: A Breakthrough Discipline for Innovation and Sustainability, Pitman Publishing in *Carnoules Declaration*, Wuppertal Institute, Germany.
9. SustainAbility with Dow Europe (1995) Who Needs It? Market Implications of Sustainable Lifestyles in *Carnoules Declaration*, Wuppertal Institute, Germany.
10. Anon (1998) *Automatic identification and sorting of plastics from different waste steams—A Status report*, APME, Brussels.
11. Anon *Assessing the potential for post-use plastics waste recycling—predicting recovery in 2001 and 2006—Summary report*, APME, Brussels.
12. Brandrup, J. (1997) *Ecological and Economical Aspects of Polymer Recycling*, (VKE), 38th Microsymposium on Macromolecules, Prague, July 1997.
13. Brandrup, J. *et al.* (eds) (1995) *Recycling and Recovery of Plastics*, Hanser Publishers, Munchen, Germany.
14. Forgac, J.M. (1996) *Feedstock Recycling of Mixed Plastics*, Society of Plastic Engineers, ARC '96 Conference, Chicago, USA.
15. Dent, I.S. and Hardman, S. (1996) Plastics waste recycling—an alternative approach, *IChemE Environment Protection Bulletin*, Issue 044.

Appendix: Analysis of polymer usage for plastics packaging in Western Europe

C. Cross

A.1 Introduction

This appendix reviews the total demand for packaging polymers in the seven major economies of Western Europe: Belgium, France, Germany, Italy, the Netherlands, Spain and the UK. This total is then split between polymers used for industrial packaging applications and those used for the packaging of consumer goods. Within the volumes used for consumer goods packaging, there is then a more detailed analysis of polymer usage according to the principal packaging types:

- flexible packaging;
- thin-walled plastic containers;
- blown plastic containers; and
- caps and closures.

The volumes of plastic polymers used for liquid carton systems are also briefly considered.

A.2 Trends in total demand for packaging polymers in all Western Europe

Overall demand for polymers in all Western European countries has been consistently buoyant in recent years. This trend is set to continue, with no significant change in the rate of growth (around 3%) during the period 2000–2004 (Figure A.1). The seven leading Western European countries account for 85% of demand. No change is anticipated in this situation.

In the industrial sector, the single most significant market is now (and will continue to be) that of food, drink and tobacco, accounting for 30% of demand. Other significant industries include chemicals, motor vehicle components and paper and packaging. The majority of these industrial markets are forecast to grow steadily at between 2 and 3% per annum. The major markets in the consumer durables sector are ceramics, glass and household appliances. The single most significant market within the consumer goods sector is that of drinks (Figure A.2), with a 38% share of total polymer demand. Non-foods account for 24% of demand, with fresh foods and processed foods at 21 and

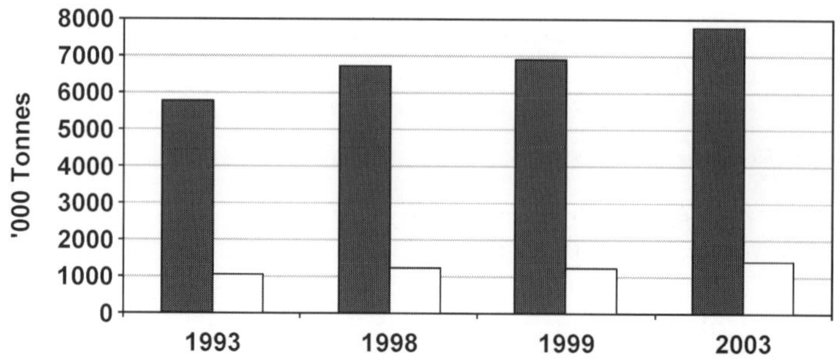

Figure A.1 Trends in packaging polymer demand. Key: ■ 7 countries; □ Other EU countries.

17%, respectively. Furthermore, the best growth prospects over the next four years will be found in the drinks market.

A.2.1 Drinks

There are several factors that continue to ensure that this market experiences healthy growth. The most significant is the trend of pack substitution from glass bottles to plastic bottles, particularly in the carbonated soft drinks, fruit juice and mineral water markets. Other factors include:

- a switch from glass bottles and liquid packaging cartons to HDPE bottles in the milk market (apparent in the UK with increasing sales of milk from supermarkets rather than doorstep delivery);
- the introduction of returnable blown plastic bottles, which are replacing returnable glass bottles in the mineral water, carbonated soft drinks and milk markets in Belgium, Germany and the Netherlands; and
- an overall decrease in the usage of glass bottles as progress in technical properties of polymers make it possible to use plastic instead of glass. For example, PET is being used for carbonated soft drinks and sparkling mineral water, and HDPE, multilayer combination and PET is being used for fruit juice;
- the increased availability of PET blown plastic bottles; and
- a wider range of bottle sizes: 500 ml, 2 l, 5 l, etc.

The beer and cider subsector is currently a mature market for polymers, but there have been numerous experiments with the use of PET for beer bottles throughout Europe; including coated PET and PET/PEN blends, for example. If the early trials prove successful, this may require a modification to our forecast of polymer demand in the beer sector.

However, all subsectors will experience good growth over the next few years. The strongest growth will be in the fruit juices sector at 10% per annum. Overall demand for polymers in the drinks market is forecast to increase by an average of 4.6% per annum to 2003.

A.2.2 Non-foods

This is thc lcast buoyant consumer goods market, with several static or mature subsectors. Overall, a growth rate of just over 1% is forecast between 1999 and 2003.

Pharmaceuticals is the largest submarket with an 18% share, followed by cleaning materials (14%) and detergents (13%). While cleaning materials and detergents will experience very little growth in polymer usage, pharmaceuticals and toiletries offer slightly better prospects at more than 1.5% per annum over the forecast period.

A.2.3 Fresh foods

The fresh foods market is forecast to grow by an average 2.3% per annum between 1999 and 2003. The dairy products sector is the largest submarket, with 58% of polymer tonnage. Growth is forecast at 2.1% per annum from 1999 to 2003, with very little competition, particularly in the dairy dessert and yoghurt end-market from non-plastics based packaging.

A.2.4 Processed foods

The two major submarkets for polymers in this sector are frozen foods (a 19% share) and chocolate and sugar confectionery (16% share). Both these markets are forecast to remain relatively buoyant, with increases over the forecast period estimated at 3.1 and 2.2% per annum, respectively.

The best growth prospects, however, will come from some of the smaller end-use markets for polymers. Heat-processed foods, pasta and rice, and sauces are all—albeit from a low base—forecast to increase at over 4% per annum between 1999 and 2003. This increase reflects the increasing popularity of ready

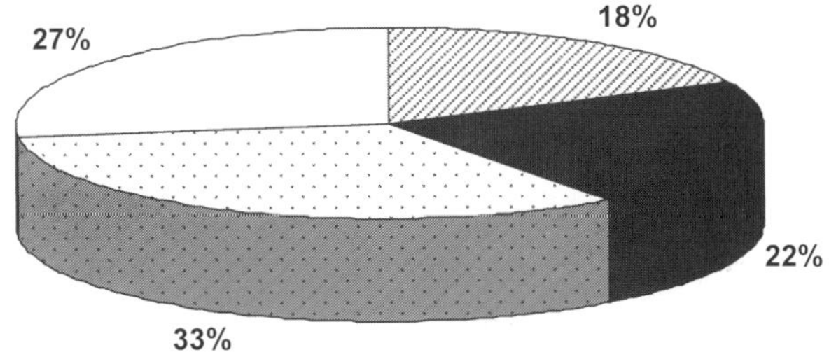

Figure A.2 Polymer demand by markets in 1998. Key: ▨ processed foods; ■ fresh foods; ⬚ drinks; □ non-foods.

meals (including those heat-processed meals packed in trays and thin-walled containers) and the popularity of rice and pasta dishes.

A.3 Total polymer demand in consumer goods markets—Western European countries

PE, PET and PP will continue to predominate in consumer goods markets throughout the forecast period (Figure A.3). However, the market share held by PE is gradually decreasing from 33% in 1993 to a forecast 29% in 2003. Conversely, the share held by PET has risen quite substantially from 20% in 1993 to a forecast 31% in 2003. The main reasons for growth in demand for PET are:

- a switch from PVC to PET bottles in the drinks and edible oil markets;
- a switch from glass bottles to PET bottles, particularly in the soft drinks and mineral water markets;
- the increased availability of PET at lower prices which has resulted in PET being used in an increasing number of end-markets that are not traditional 'PET' users, for example cleaning materials, detergents, chocolate and sugar confectionery (inserts) and frozen foods (thin-walled containers); and
- the introduction of new PET pack types, replacing glass containers such as PET bottles for honey or PET containers for dry beverages.

Over the next few years, PET will experience a higher annual increase (5.5%) than will PE (1.5%).

PP tonnage has been expanding at a steady rate and over the next four years the growth rate will begin to level out at just over 3% per annum. The growth

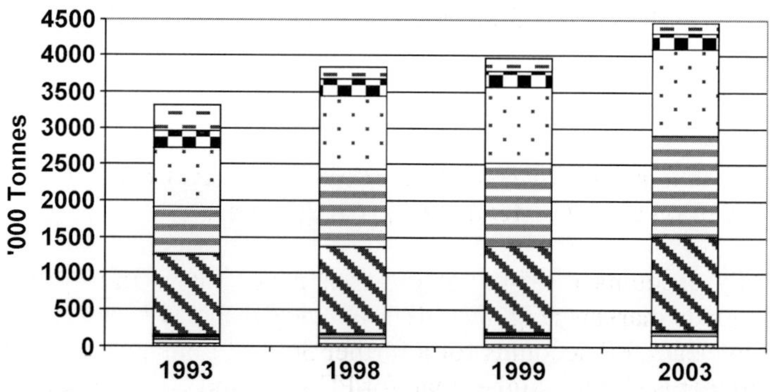

Figure A.3 Packaging polymer demand for consumer goods markets. Key: □ multilayer; ▨ PE; ▤ PET; ⬚ PP; ▪ PS; ▤ PVC.

in demand for PP can be attributed to several factors:

- the large scale substitution of PS by PP in the dairy markets, with most significant changes in cheese, cream, dairy desserts and yoghurts;
- the switch from HDPE to PP for closures, as PP is better adapted for speciality closures; and
- the increasing usage of PP bottles, in particular for non-foods markets such as those for cosmetics, toiletries, haircare products, cleaning materials and pharmaceuticals.

PVC tonnage will continue to decline but at a slower rate than in the previous five years (−5.8% per annum between 1999 and 2003, compared with −11.9% between 1992 and 1997). The two main reasons for the decline in demand are: a) concerns that PVC waste was potentially hazardous; and b) the boom in demand for PET for its clarity and competitive cost, offering an attractive substitute to PVC.

Demand for PS is also declining, at a slower rate than for PVC. PS accounted for 8% of demand five years ago but by 2003 this polymer will account for under 5% of tonnage. PS usage has been concentrated in the dairy markets, where it has been competing with PP. PP is now preferred to PS in an increasing number of applications, for environmental and technical reasons.

Other polymers, including PA, PC and multilayer, are forecast to rise in usage. Multilayer polymers have increased in usage in consumer goods markets; while their share of the market will remain at around 2.5 to 3%, demand for multilayer is forecast to grow by over 8% per annum to 2003. Multilayer combinations are increasingly popular for blown plastic bottles and thin-walled containers in the food and drinks markets. Multilayer thin-walled containers are traditionally used for fresh meat, fish and poultry and for heat-processed foods (ready meals and fruits). Multilayer bottles are largely used for sauces, as multilayers provide the necessary barriers to gases and flavours. They are also being introduced for fruit juice and carbonated soft drinks, where they are expected to make significant inroads.

A.4 Polymer demand for flexible packaging in consumer goods markets—Western European countries

PE currently accounts for 54% of polymer usage for flexible packaging in consumer goods markets (Figure A.4). Demand for PE will remain stable over the next few years. PP accounts for a further 36% of tonnage. The growth rate for this polymer is slightly higher than for PE at around 2% per annum to 2003. Among other polymers used, PET is expected to have the strongest growth at 2.8% per annum. PVC usage has been static overall during the last five years'

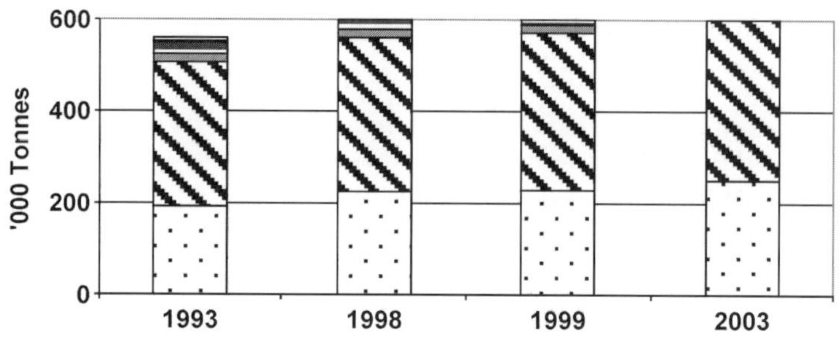

Figure A.4 Polymer demand for flexible packaging in consumer goods markets. Key: ⬚ PP; ▧ PE; ▩ other polymers; □ multilayer; ■ PA.

and will experience a relatively small recovery in demand (up 1.7% per annum) over the forecast period.

A.4.1 PE

In the fresh foods market, PE dominates with a 58% share of consumption. PE tonnage in this sector has been decreasing at a rate of 1.6% per annum over the past five years. This decrease can be linked to the declining polymer tonnage in the cheese market (owing to pack weight reductions) and to the substitution of PP in the bread and bakery goods market. In the forecast period, a decrease of PE in the bread and bakery goods market will be counterbalanced by rising PE consumption in the fresh meat, fish and poultry markets.

Within the processed foods sector, the main markets for PE are breakfast cereals, dry beverages, frozen foods, and rice and pasta. Frozen foods represent 10% of the overall PE tonnage. PE demand in this sector is expected to show healthy growth over the forecast period. Non-foods is a smaller end-market for PE with demand being limited to a few applications such as paper products, pet foods and detergents. PE is also used for flexible packaging for milk in Northern Europe.

A.4.2 PP

The processed foods markets represent the main end-market for PP films (68% of the total PP tonnage for flexible packaging). Chocolate confectionery, biscuits and cakes and snack foods are the three main consumers of PP and in these markets, growth will remain relatively stable during the forecast period. OPP films are considered well adapted to high filling speeds on flow-pack machines; they offer good barrier properties against humidity and a good printing surface. Fresh foods markets represent less than 23% of PP consumption, demand being concentrated in the fresh fruit and vegetables, and bread and bakery goods markets. Non-foods markets are less significant with 9% of the tonnage.

A.4.3 Other polymers

Demand for PET is increasing in a number of markets. An above average increase is forecast in demand for PET in the frozen foods (+10% per annum) and pasta and rice sectors (up 7% per annum). Demand for PET is also growing in several fresh food sectors, for example, bread and bakery goods, and meat products.

A.5 Polymer demand for thin-walled plastic containers in consumer goods markets—Western European countries

Thin-walled plastic containers include dairy products containers, insert trays for chocolate and sugar confectionery, biscuits and cakes and thin-walled trays for fresh meat products. PP and PS are the dominant polymers, with PP accounting for 53% of demand and PS 25%. However, demand for PS is in decline whereas prospects for PP remain very healthy (Figure A.5).

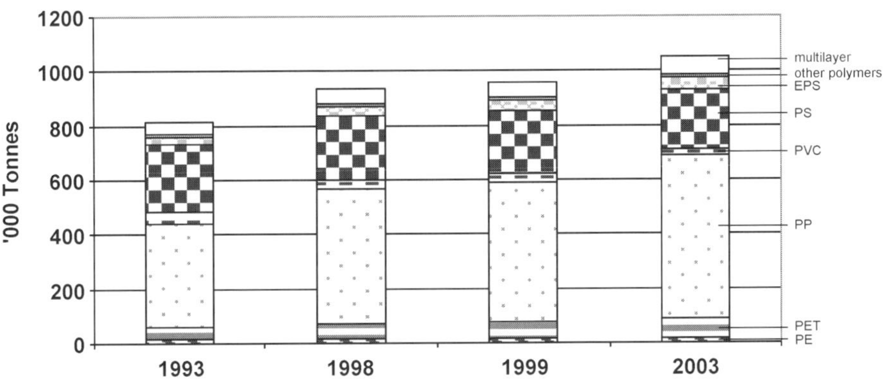

Figure A.5 Polymer demand for thin-walled containers in consumer goods markets.

A.5.1 PP

Fresh foods (dairy products and other fresh foods) is the main sector for PP, accounting for 60% of the total PP tonnage. PP has been growing at a fast rate in this sector, increasingly replacing PS in a number of applications. The markets where substitution has been most pronounced are yoghurt, dairy desserts, cheese, fresh fruit and vegetables. Substitution is occurring as a result of several factors:

- Increased production of value-added products (in particular for dairy desserts and yoghurts), which generally come in PP containers, as more fancy shapes are possible;
- PS is not considered rigid enough for some applications (fresh fruit and vegetables);
- PP prices have become more competitive in recent years; and
- developments in PP polymer and in processing equipment.

Application of PP for processed foods represent 31% of the total PP tonnage, with frozen foods and butter and fats being the two biggest users. PP tonnage has increased significantly in these markets as a result of the substitution of PP for PS. The reasons are:

- better technical properties of PP, such as barrier and rigidity; and
- PP is considered more cost effective.

In the biscuits and cakes, and chocolate and sugar confectionery markets, where PP is used for insert trays, the share taken by PP has been increasing to the detriment of PVC. Non-foods account for the remaining 9%, cosmetics, paints, and stains and varnishes being the two biggest end-markets. PP tonnage in the non-food segment is forecast to increase at a rate of over 4% per annum for the following reasons:

- pack substitution in the paints markets;
- increase in overall production levels; and
- PP tends to replace PVC or PS in a few applications.

A.5.2 PS

Overall, PS tonnage is rapidly declining. The dairy sector, especially yoghurt and cheese, remains the main market for PS, accounting for 78% of the total PS tonnage. Although PS is losing out to PP in all fresh foods/dairy products markets, it remains largely in use for form-fill-seal applications. Within the fresh food sectors, PS is still holding an important share in cheese, dairy desserts, and fresh fruit and vegetables. Usage of PS within processed foods is most significant in frozen foods (7.4% of the total PS tonnage), chocolate and sugar confectionery (2.6%), and butter and fats. PS is declining rapidly in the processed food markets, except where it is used as insert trays. Non-foods is a small market for PS with just over 3% of the total PS tonnage, being concentrated in the cosmetics and pharmaceuticals sectors.

A.5.3 Other polymers

PET currently accounts for 6% of tonnage; demand is expected to increase at an average of 4.7% per annum between 1999 and 2003. In the processed foods sector, while PET is used in many markets, its main application is trays for chilled and frozen ready meals, representing 24% of total PET tonnage. It is in the frozen food markets that the best prospects for PET will be found. Multilayer applications are used for heat-processed foods, fresh meat, and fish and poultry. Usage of these combinations is expected to show a healthy growth rate of 4% per annum to 2003.

A.6 Polymer demand for blown plastic containers in consumer goods markets—Western European countries

Blown plastic bottles remain the fastest growth market for polymers. PET accounts for 56% of current polymer demand and is one of the fastest growing polymers. In 1993, it accounted for only 39% of all polymers specified. HDPE usage is the second largest tonnage and represents 30% of total demand. PVC remains the third most important polymer but its share of total tonnage dropped from 20% in 1992 to 8.4% in 1997. PP accounts for a small percentage of the total tonnage (2.8%) but its share is increasing at a fast rate. Multilayer combinations are expected to be the fastest growing polymers in the 1998–2003 period, although they will still only account for some 27% of demand (Figure A.6).

A.6.1 HDPE

HDPE usage is significant for drinks, particularly milk, and for non-food markets. Milk accounts for 35% of the total HDPE tonnage. It has been an increasing sector for HDPE as blown plastic bottles have been developing at the expense of glass and liquid packaging cartons. Non-food markets represent 61% of the HDPE tonnage. Cleaning materials and detergents are important end-markets for HDPE, together representing 40% of the tonnage. By 2003, this will have decreased to 37.5% owing to limited polymer substitution and overall tonnage reduction, as bottles are becoming lighter. Other important markets for HDPE

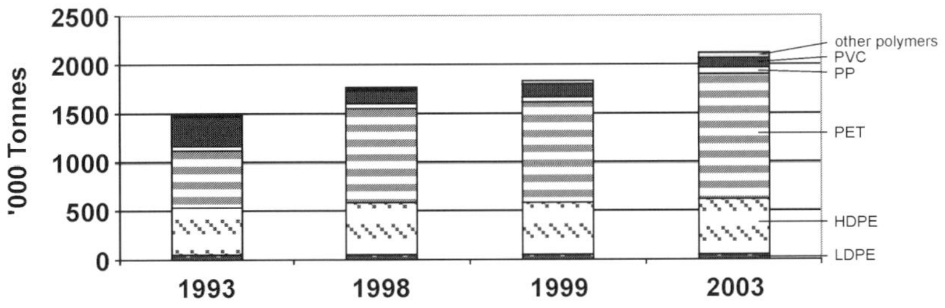

Figure A.6 Polymer demand for blown plastic containers for consumer goods markets.

include haircare products, toiletries and automotive products, where overall usage is expected to remain stable, although a slight decline is expected for automotive products as a result of declining production levels. In the food markets, HDPE is used for edible oils packaging. HDPE usage is also expanding at a fast rate (albeit from a small base) for yoghurt, as liquid yoghurt is becoming increasingly popular.

A.6.2 PET

The drinks industry accounts for 91% of PET demand, the two main end-markets being mineral water and carbonated soft drinks, accounting for 88% of the total tonnage. PET usage in these two markets has been growing, as a result of the following factors:

- the switch from PVC to PET by all main mineral water producers;
- the switch from glass bottles to PET bottles for both mineral water and carbonated soft drinks; and
- the increasing production levels.

Non-food markets account for 4.7% of PET demand, the two most significant markets being cleaning materials and detergents, where PET is used instead of PVC when clarity is required. Other important markets are personal care, including haircare products, toiletries and cosmetics, where PET is found for upmarket products, generally replacing glass containers and, to a lesser extent, PVC. Processed food accounts for 4% of total PET demand with most of the tonnage found in edible oils, where a move from PVC to PET is taking place. PET is also gaining importance in the sauces market.

A.6.3 PP

PP has been gaining ground in all end-markets. Non-foods currently account for 80% of PP tonnage. The markets where PP is significant are cleaning materials, haircare products, toiletries and cosmetics, with personal care markets offering the best growth prospects. Processed foods currently account for the remaining 20%, with sauces being the main end-market. Usage of PP in the drinks market is forecast to grow rapidly following its introduction for milk in the UK.

A.6.4 Other polymers

Multilayer combinations are used for food and drinks, as they provide the necessary barriers. Sauces, fruit juice and carbonated soft drinks are the main multilayer users. Other polymers include mainly PC, found in returnable bottles for milk and liquid yoghurt in the Netherlands and Germany.

A.7 PE demand for liquid packaging cartons in consumer goods markets—Western European countries

Demand for PE for liquid packaging cartons has remained relatively constant over recent years (Figure A.7), and no significant changes are expected for the next four years with the growth rate remaining at around 1.7% per annum. PE demand for liquid packaging cartons amounts to over 130 000 tonnes and is expected to increase by 1.7% per annum over the next four years.

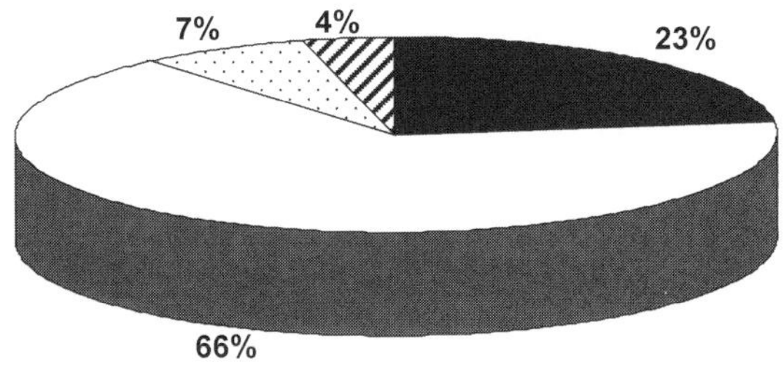

Figure A.7 PE demand for liquid packaging cartons by markets in 1998. Key: ■ fruit juice; □ milk; ⬚ other drinks; ▨ foods.

The drinks sectors account for 93% of PE demand for liquid packaging cartons. Milk alone accounts for 61.2% of PE consumption. Milk is a mature market, where no significant increase in liquid packaging cartons is taking place. PE tonnage is therefore expected to remain stable (the growth of PE in milk packaging is in blowmoulded containers). Fruit juice accounts for 23.3% of PE demand. It is a dynamic market where PE is forecast to grow by 3.9% over the next four years. Other markets include non-carbonated soft drinks (5.2% of the tonnage)—where PE demand is also on the increase—and wines and spirits (2.7%), where PE consumption is expected to increase by 3% per annum.

Usage of liquid packaging cartons is limited in other markets. Dairy products account for 3.9% of the PE consumption, mainly found for cream and, to a lesser extent, for dairy desserts and yoghurts. Tonnage of PE for dairy products is forecast to remain stable. Processed foods represents 2% of the PE tonnage; PE consumption mainly taking place in the heat processed foods market, and to a negligible extent for edible oils. Consumption of PE for detergents has been growing at a very rapid rate (+17%); this will now slow to around 5%. However, current PE demand in this market represents only 0.3% of total tonnage.

A.8 Polymer demand for plastic closures in consumer goods markets—Western European countries

The tonnage of polymers used for closures reached 400 386 tonnes in 1998 and is growing in accordance with the increasing usage of plastic bottles and other

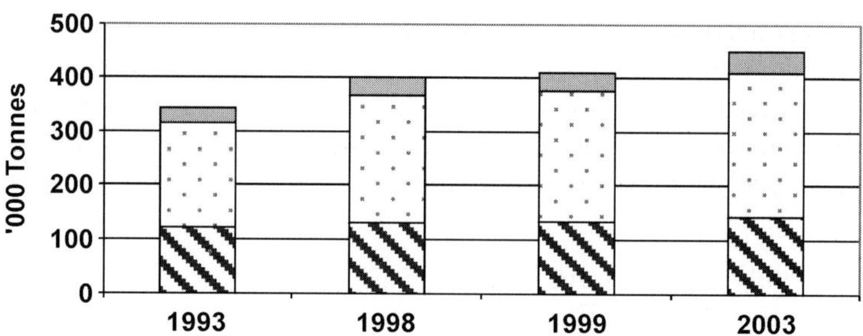

Figure A.8 Polymer demand for plastic closures for consumer goods markets. Key: ▨ PE; ⬚ PP; ▩ other polymers.

plastic-based pack types (Figure A.8). Also, plastic closures are increasingly being fitted on glass containers. HDPE and PP dominate. PP is by far the main polymer to be used with a 59% share of total tonnage. HDPE accounts for 31%. Other polymers include LDPE, ABS and PETG.

A.8.1 PP

Non-food markets account for 63% of PP demand. Within non-foods, demand is concentrated in the personal care markets, cosmetics, toiletries and pharmaceuticals. Other significant markets include haircare products, detergents and cleaning materials. PP demand is growing steadily in these markets as a move from HDPE to PP is taking place. PP, being a better material for speciality closures, is increasingly popular.

Drinks represent 19% of PP demand, carbonated soft drinks being the main end-market with 14% of the total demand. The rigidity of PP makes it a suitable material for this type of closure. In all drinks markets, there is a slow move away from HDPE towards PP, the exception being mineral water. Overall, PP demand for drinks closures is forecast to grow at a rate of 4% per annum over the next four years. Processed foods account for the remaining 18% of the PP demand for closures, the main market being dry beverages (12.5% of the total tonnage). Jams and preserves and sauces are also important. A switch to PP is also taking place in these markets.

A.8.2 HDPE

Drinks is the main sector for HDPE, accounting for 50% of total demand, with mineral water alone accounting for 34%. HDPE is the only polymer used for mineral water closures in a number of countries, as PP usage is limited to sparkling water. Other significant markets include milk and carbonated soft drinks. HDPE tonnage for drinks is growing at a fast rate, mainly because of the growth in sales of mineral water.

Non-foods account for 41% of tonnage. HDPE usage is widespread in all non-foods markets but automotive products is the only market where HDPE clearly dominates. The most important market is pharmaceuticals. Foods represent the remaining 9%, edible oils being the most significant. The main use for LDPE in packaging is for milk in the UK.

A.9 Total polymer demand in industrial markets—Western European countries

Polymer demand in industrial markets has been growing by just over 3% per annum over the past five years. Over the next few years, demand is expected to remain relatively stable to reach over 3 901 000 tonnes in 2003; the per annum average growth rate slowing only very marginally to 2.8% between 1999 and 2003.

In industrial markets PE predominates, accounting for 86% of total polymer demand. No change is expected in the share of the market held by PE over the forecast period (Figure A.9).

A.10 Polymer demand for flexible transit packaging in industrial markets—Western European countries

Throughout the forecast period, PP shrink and stretch film will continue to account for 90% of demand for polymers for flexible transit packaging. The share of this sector held by PP is expected to drop slightly to 6.9% by 2003 as there is some substitution of PP by other polymers apparent in the market (Figure A.10).

A.11 Polymer demand for rigid plastics containers in industrial markets—Western European countries

Currently, PE accounts for 79% of total polymer tonnage and PP 17%. By 2003, the share held by PE will have decreased slightly to 78% (Figure A.11). The

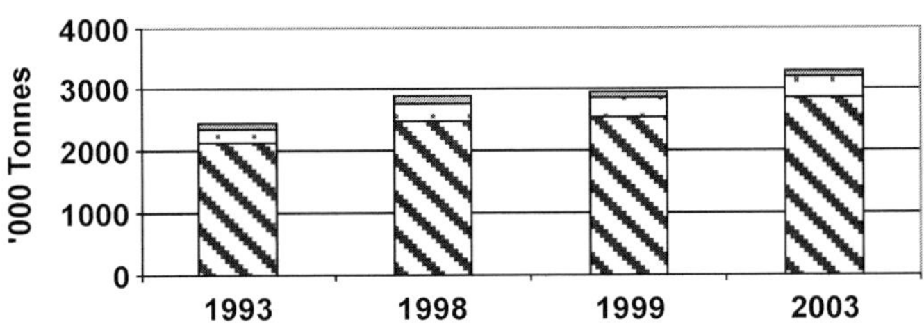

Figure A.9 Packaging polymer demand for industrial markets. Key: PE; PP; other polymers.

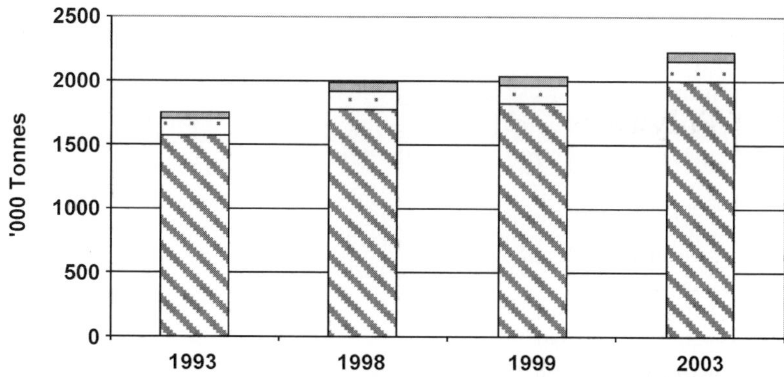

Figure A.10 Polymer demand for flexible transit packaging. Key: PE; PP; other polymers.

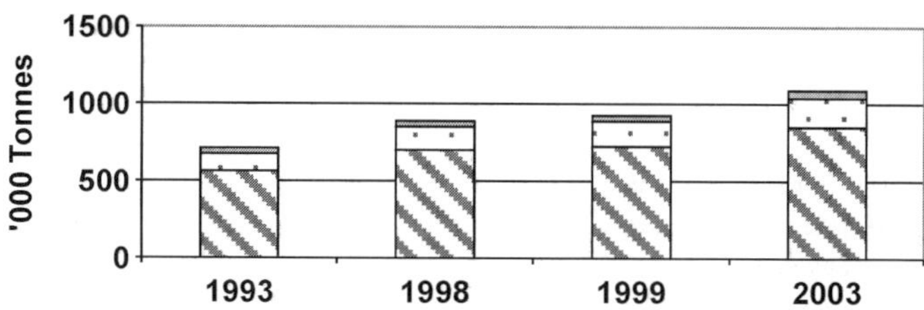

Figure A.11 Polymer demand for rigid plastic containers. Key: PE; PP; other polymers.

chemical industry, food, drink and tobacco, and motor vehicle and components are the main areas where PE is specified for rigid plastic containers. Food, drink and tobacco are the main markets for PP rigid plastic containers.

List of Polymers

ABS	Acrylonitrile/Butadiene/Styrene
Acetal	Polyoxymethylene
Acrylic	Polymethylmethacrylate
EVA	Ethylene/Vinyl Acetate
EVOH	Ethylene/Vinyl Alcohol
HDPE	High Density Polyethylene
LDPE	Low Density Polyethylene
LLDPE	Linear Low Density Polyethylene
PA	Polyamides e.g. PA6, PA12 (Nylons)
PBT	Polybutylene Terephthalate
PC	Polycarbonate
PE	Polyethylene
PEN	Polyethylene Naphthalate
PET	Polyethylene Terephthalate
PP	Polypropylene
PS	Polystyrene
PVC	Polyvinyl Chloride
PVDC	Polyvinylidene Dichloride
TPE	Thermoplastic Elastomer
UPVC	Unplasticised PVC

Index